(*continued on back*)

A Handbook of
Introductory Statistical Methods

A Handbook of Introductory Statistical Methods

C. PHILIP COX

Iowa State University

JOHN WILEY & SONS

New York • Chichester • Brisbane • Toronto • Singapore

Library of Congress Cataloging in Publication Data:

Cox, C. Philip (Charles Philip),
   A handbook of introductory statistical methods.

   (Wiley series in probability and mathematical
statistics. Applied probability and statistics,
ISSN 0271-6356)
   Bibliography: p.
   Includes index.
   1. Statistics. I. Title. II. Series.

QA276.12.C69 1987   001.4'22   86-13137
ISBN 0-471-81971-9

Printed in the United States of America

10 9 8 7 6 5 4 3 2 1

To Springfield, A. B. G. S., Aedes Christi, Frank Yates,
Ted Bancroft, and the perspicacious charmer

Preface

I have attempted to give here an economical summary of some commonly used statistical analysis procedures and their basic principles. Regarding the statistical component as the infrastructure of heuristic and descriptive data analysis, I emphasize methodological aspects to complement the instruction in a first course for graduate students who intend to use statistical methods in their own research. Experience has suggested that the comprehension of ideas in class is often attenuated by, in anticipation of *not* understanding, compulsive note-taking. Experience has shown that students can rid themselves of this particular didactic tyranny and enhance *understanding* and the efficiency of the whole information transfer process, once they become confident that the essential procedural details and formulas are summarily available. Additionally, it is hoped that this handbook will provide a convenient reference source of statistical procedures for practicing scientists in quantitative research disciplines.

To reduce another common impediment, the commission, diagnosis, and eradication of unorthodox arithmetic, unveridically simple data are often used in illustrative examples. This is to allow accurate computational "mechanics" to be learned quickly and to furnish time for the development of intellectual statistical analysis skills at which computers are, as yet, incompletely proficient. No calculus is used.

Most of the methods described rely on the assumption of Gaussian distributed observations, and uncertainly valid assumptions impose qualifications on the reliable applicability of statistical methods. The strengths and weaknesses of these and their possible applications to nonregular situations are better appreciated when their bases are well understood, which is why some derivations are outlined. These offerings, for those also interested in the "Why?", generally appear in sections marked "+" which can be skipped unconcernedly by students wishing to concentrate on the "How?". Also, it may be noted that, for convenient reference, all formulas, figures, and tables are serially ordered within sections, thus Eq. (8.5) is the fifth equation in Section 8 and Table 40.2 is the second table in Section 40.

I shall much appreciate suggestions for improvement and I am already indebted to academic generations of my "Stat-401" students at Iowa State University for suggestions cumulatively incorporated. I also thank these students for encouraging this publication, not least by their persistently adverse comments on my handwriting. Particularly in this, but in un-counted other respects, several secretaries have made inestimable con-tributions including that elusive art which bestows aesthetic merit on the presentation of even statistical formulas and statements. It is a special pleasure to recognize Donna L. Nelson's contribution here—the produc-tion of the final draft with apparently tireless cheerfulness and certainly impressive competence. For special contributions I also wish to thank another *sine qua non*, Wiley-Interscience editor Beatrice Shube. Although sins of omission and commission are mine, I am indebted to several professional colleagues and especially thank Dr. Fred Lorenz and Dr. Edward Pollak for reviews of particular sections. Other colleagues have contributed indirectly; I know that most of them appreciate that prefaces should be advisedly truncated.

C. PHILIP COX

Ames, Iowa
September 1986

The Pyrgus: An Early Randomizing Device

The wooden dice box illustrated below was discovered in November 1931 by L. P. Kirwan during excavations of covered tombs attributed to "The X-Group," a culture on the Nubian Nile (ca. 300–400 A.D.). A Roman calendar for 354 A.D. illustrated a similar dice box.

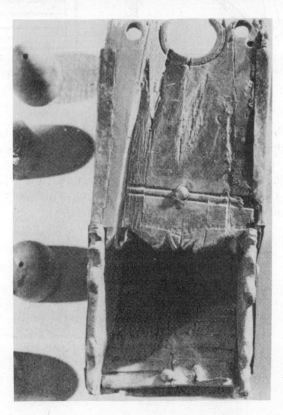

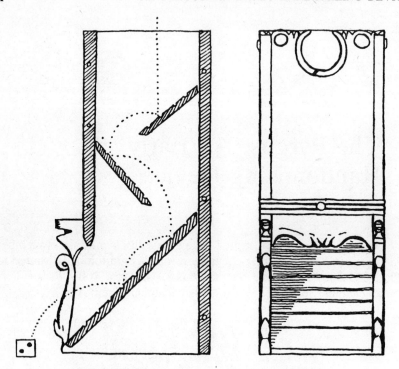

Cross section and frontal view of the "cheat-proof" dice box.

"You dropped the dice in through the top and they bounced down over a ladderlike arrangement of boards and out through an opening, flanked by carved dolphin heads, as a safeguard against cheating."

Reproduced, by kind permission, from L. P. Kirwan, A Little-Known People of the Nubian Nile, in E. Bacon (Ed.), *Vanished Civilizations of the Ancient World*, McGraw-Hill, New York, 1963.

Contents

Notes and Notations

For Formulaphobes: Algebraic symbols and formulas are convenient forms for whole families of individual numerical quantities. Mathematics is the language of these forms. The forms, as symbols and groups of symbols, are the nouns, subjects, and objects in, as equations and inequalities, the sentences. Derivations change, develop, and reveal connections between forms. Please think of the family forms as entities in themselves and identify them carefully; give them their names in their contexts. The sentences can then be read and their stories unraveled. Some common statistics forms follow.

Greek Letters (Pronunciations): α (alfa); β (bayta); δ (delta); ε (epsilon); χ (kie); μ (mew); π (pie); ρ (row); Σ (sigma); σ (sigma); ζ (zeeta).

Useful Conventions: (i) Roman letters for random variates (such as estimates), Greek letters for fixed population parameters. Examples: x and $\bar{x}$ estimate the parameter μ_x; r is the estimate of the parameter ρ; s^2 estimates the parameter σ^2. (ii) A circumflex or "hat" denotes an estimate of the hatted parameter; thus $\bar{x}$, an estimate of μ_x, is a $\hat{\mu}_x$, and s_x^2 is a $\hat{\sigma}_x^2$.

Symbolism: $|x|$ = absolute value of x, (6.2); $\sum_{i=1}^{i=n} x_i = (x_1 + x_2 + \cdots + x_n)$, (2.1); $\sum' xx = \sum_1^n (x_i - \bar{x})^2$, (2.5); $\sum' xy = \sum_1^n (x_i - \bar{x})(y_i - \bar{y})$, (48.1); $\leq$, "not greater than." $\mu[y] = \mu_y = E(y)$ = the population mean of the random variate y, whatever y is, (§5). $\text{Var}(y) = V(y) = \sigma_y^2$ = the variance of the population of which y is a random member, (5.1). Thus, with $y = (x - \mu_x)^2$, alternative notations for population variance are $\mu[(x - \mu_x)^2] = \text{Var}(x) = V(x) = \sigma_x^2$.

"$\sim$" and "$,\ldots,$" can be read as "is distributed in the pattern of" and "and so on up to," respectively. Thus, in §13, $x_i \sim NI(\mu_x, \sigma_x^2)$, $i =$

$1, 2, \ldots, n$, is read as "the random variates x_1, x_2, and so on up to x_n are distributed in the normal distribution pattern and are independent of each other." The population mean of their normal distribution is μ_x; the population variance is σ_x^2.

$P[A] = p$ means "the probability that the statement 'A', within the [ ], is true is equal to the positive number p where $0 \leq p \leq 1$." Probability statements are made about random variates (*not* about constants). Thus, $P[|t_{10}| > 2.228] = 0.05$ is the probability that the absolute value of a randomly selected 10-degrees-of-freedom t-variate will exceed 2.228; see §4 and (12.3). An integer subscript to a variate name denotes the number of degrees of freedom. However, the subscript can be less than 1, for example, v_α, with $0 < \alpha < 1$, denotes that value of the variate v, for which $P[v > v_\alpha] = \alpha$, so that the subscript α is the area in the right-tail, beyond the point v_α, of the v-distribution. Thus, for the χ_{10}^2-variate in §15, $\chi_{0.025}^2 = 20.843$, means that $P[$a random χ^2, 10-degrees-of-freedom, variate $> 20.843] = 0.025$. When both degrees of freedom and right-tail areas are to be shown, a notation such as $\chi^2(10; 0.025) = 20.843$ is useful.

A vertical stroke, $|$, signifies conditionality; that is, some restriction of the general situation. Thus, $\sigma_{y|x}^2$ in simple linear regression (§47), denotes the population variance of y-values conceptually observable at any one, supposedly fixed, x-value. The stroke helps the multiple regression notation in §127 where β_1, for example, is the abbreviation for $\beta_{y1|2\cdots p}$, the regression coefficient of y on x_1, when $x_2, \ldots, x_p$ are constant.

The R-Notation: $R(\mu, \beta_1, \beta_2)$, for example, denotes the 3df sum of squares "due to" the three parameters μ, β_1, and β_2 for the multiple regression model (127.6) with $p = 2$. $R(\beta_3, \beta_4 \mid \mu, \beta_1, \beta_2)$ denotes the additional sum of squares due to the inclusion of β_3 and β_4, the parameters preceding the $|$, to augment a model that already includes the parameters, μ, β_1, and β_2, which follow the $|$.

Reference Distributions

Variate	Analysis Uses	Definition Section	Table
z	for means, σ^2 known	§11	A2
$t(f)$	for means, σ^2 unknown	§12	A3
$\chi^2(f)$	for variances and for discrete data	§14	A4
$F(f_1, f_2)$	for variance ratios	§26	A5
Binomial	in specific situations	§31	—
Poisson	in specific situations	§44	—
$q(a, f)$	for multiple range tests	§100	A6

How Many Decimal Places?: The Roman emperor Agrippa (63–12 B.C.) made a very early interval estimate, one for the dauntingly irregular and stochastic perimeter of Britain. Lloyd (1966) reports on this as: "Great store was set, in early times, on computing its circuit, which Agrippa, decorating a very broad margin with delicate precision, declared to be not more than 28,104 stadia, nor less than 20,526 stadia—a stadium being about 202 yards." Hand electric calculating devices are very generous with decimal places and no rounding is essential in intervening calculations. For common sense final reporting of estimates and intervals, however, rounding to two effective digits, that is, to two digits other than zero, is recommended for convenience in general practice—as advocated by Ehrenburg (1981).

A Handbook of
Introductory Statistical Methods

Some Basic Concepts
and Procedures

§1. DEFINITIONS

Repeated observations, which exhibit or are subject to variability when taken under the same circumstances, are regarded as samples of particular values, measurements or realizations, of a *statistical variate*.

A **continuous variate** is one which can conceptually take any value within its naturally occurring range. Examples are battery life, blood pressure, and cholesterol concentration.

A **discrete variate** is one for which the possible values are separated by finite intervals. Examples are heart rate, family size, and radiation count.

The individual variate values in a sample are regarded as particular members drawn from a conceptual statistical "parent" population.

A **statistical population** is the aggregate of all possible variate values, whether different or not, conceivably obtainable as single observations in the specific experimental circumstances. In this concept, repeated, identical, variate values are regarded as different members in the population.

A **sample** is that set which consists of just those values, from the aggregate or parent population, that are *actually obtained as observations*.

A **random sample** is obtained by taking observations independently of each other to ensure that *each member* of the population has an *equal chance* of being selected, that is, of appearing in the sample.

Independence: Observations, or values of variates, are pairwise independent when the value of any one is in no way related to, predictable from, or restricted by the value of any other.

1

In any enquiry it is important to be specific about the appropriate parent population and to ensure that observations are random and independent members of just this population.

Indices of scientifically interesting characteristics of populations are known as **parameters**. Important among these are the following:

(a) Location parameters which locate the population in terms of the distance from the origin of measurement to some defined feature of the population. Examples are:

   (i) **the population mean**: the average of all the variate values;

   (ii) **the population median**: half of all the variate values exceed and half are less than the median;

   (iii) **the population mode**: the magnitude of the most frequently occurring, or modal, variate value.

(b) Parameters which provide indices or numerical measures of the variability from one observation to another. Examples are:

   (iv) **the population variance**: the average of the squares of the deviations of the variate values from their population mean;

   (v) **the population standard deviation**: the square root of the population variance.

Populations may differ in their locations and in their variability, differences which would be indicated by different numerical values of the corresponding parameters. In seeking information about a population we seek information about its parameters. These could only be absolutely determined if the value of every one of the population members were known. Such knowledge is rarely achievable. Instead it is necessary to make inferences about the population parameters using their estimates calculated from random samples.

A **point estimate** is calculated to provide the best *single value* for an unknown parameter.

An **interval estimate** (e.g., Agrippa's, p. xxi) is a *range of values* within which an unknown parameter is estimated to lie.

A **confidence interval** is an interval estimate for which, *with specified confidence*, it is asserted that the value of the parameter lies in the interval.

Statistical methods use observations in relatively small samples:

   (i) to obtain point and confidence interval estimates of population parameters;

(ii) to assess the probabilities that sample findings are in some agreement, that is, are statistically compatible, with hypothesized values for the unknown population or "true" values of the parameters;

(iii) to make inferences about differences and ratios of population parameters.

§2. SAMPLE MEAN AND VARIANCE: COEFFICIENT OF VARIATION

Sample mean: If $x_1, x_2, \ldots, x_n$ are the values in a sample of n observations, the sample mean is the quantity or statistic $\bar{x}$, defined as

$$\bar{x} = \frac{1}{n}(x_1 + x_2 + \cdots + x_n) = \frac{\sum_{i=1}^{n} x_i}{n} \tag{2.1}$$

where the sum, $\sum_{i=1}^{n} x_i = (x_1 + x_2 + \cdots + x_n)$, is read as "Sigma x_i from $i = 1$ to $i = n$." The sum, $\sum_{i=1}^{n}(x_i - \bar{x})$, of deviations about the sample mean is zero because

$$\sum_{1}^{n}(x_i - \bar{x}) = \sum_{1}^{n} x_i - \sum_{1}^{n} \bar{x} = \sum_{1}^{n} x_i - n\bar{x} = (x_1 + x_2 + \cdots + x_n) - n\bar{x}$$

$$= 0 \quad \text{using (2.1)}$$

If n_j denotes the number of observations in the sample which have the same value x_j, the sum of the observations is

$$\sum_{1}^{n} x_i = \sum_{j} n_j x_j$$

where the sum of products on the right is now taken over all the *different* values, the sequence x_j say, of x. From (2.1) we then have

$$\bar{x} = \frac{\sum_{j} n_j x_j}{n} = \sum x_j \left(\frac{n_j}{n}\right) \tag{2.2}$$

The fraction n_j/n is the *relative frequency* of the observation x_j in the sample and, writing $f(x_j) = n_j/n$, (2.2) becomes

$$\bar{x} = \sum x_j f(x_j) \tag{2.3}$$

4   SOME BASIC CONCEPTS AND PROCEDURES

Furthermore, since $\sum n_j = n$, it follows that

$$\sum f(x_j) = \sum \left(\frac{n_j}{n}\right) = \frac{\sum n_j}{n} = 1$$

that is, *the sum of all the relative frequencies is unity.*

Example: Suppose that the sample observations (values of the variate) are

$$x_1 = 3.2, \quad x_2 = 2.9, \quad x_3 = 4.6, \quad x_4 = 2.8, \quad x_5 = 2.9,$$

$$x_6 = 3.2, \quad x_7 = 3.2, \quad x_8 = 2.8$$

From (2.1), with $n = 8$,

$$\bar{x} = \frac{3.2 + 2.9 + \cdots + 2.8}{8} = 3.2$$

Also, to calculate the deviations $(x_i - \bar{x})$ for $i = 1, 2, \ldots, 8$:

x_i	3.2	2.9	4.6	2.8	2.9	3.2	3.2	2.8
$\bar{x}$	3.2	3.2	3.2	3.2	3.2	3.2	3.2	3.2
$(x_i - \bar{x})$	0.0	−0.3	1.4	−0.4	−0.3	0.0	0.0	−0.4

So

$$\sum_{i=1}^{8} (x_i - \bar{x}) = 0.0 - 0.3 + 1.4 + \cdots - 0.4 = 0$$

Now rewrite the sample as

$$2.8, \quad 2.8, \quad 2.9, \quad 2.9, \quad 3.2, \quad 3.2, \quad 3.2, \quad 4.6$$

Then, as in (2.2) with the x_j sequence 2.8, 2.9, 3.2, and 4.6,

$$\bar{x} = \frac{2(2.8) + 2(2.9) + 3(3.2) + 1(4.6)}{8}$$

$$= 2.8\left(\frac{2}{8}\right) + 2.9\left(\frac{2}{8}\right) + 3.2\left(\frac{3}{8}\right) + 4.6\left(\frac{1}{8}\right) = 3.2$$

wherein the successive values of $f(x_j) = n_j/n$ are 2/8, 2/8, 3/8, and 1/8. Also, since,

$$\sum n_j = 2 + 2 + 3 + 1 = 8 = n$$

it is confirmed that

$$\sum f(x_j) = \sum \left(\frac{n_j}{n}\right) = \frac{2}{8} + \frac{2}{8} + \frac{3}{8} + \frac{1}{8} = 1$$

Sample variance: This descriptive statistic provides *a numerical measure of the variability* in a sample of observations. If the sample members are $x_1, x_2, \ldots, x_n$, the sample variance is the quantity or statistic defined as

$$s^2 = \frac{1}{n-1}\{(x_1 - \bar{x})^2 + (x_2 - \bar{x})^2 + \cdots + (x_n - \bar{x})^2\} \qquad (2.4)$$

wherein "..." is read "and so on up to." It is convenient to abbreviate the sum of squares of deviations from the mean as

$$\sum_1^n (x_i - \bar{x})^2 = (x_1 - \bar{x})^2 + \cdots + (x_n - \bar{x})^2 = {\sum}' xx \qquad (2.5)$$

so that (2.4) becomes

$$s^2 = \frac{{\sum}' xx}{n-1} \qquad (2.6)$$

As is noted later (§5), ${\sum}' xx$ is divided by $(n-1)$, rather than n, because of the role s^2 plays as an estimator of the population variance.

Sample variance computation: Since

$$(x - \bar{x})^2 = x(x - \bar{x}) - \bar{x}(x - \bar{x})$$

it follows that

$${\sum}' xx = \sum x(x - \bar{x}) - \sum \bar{x}(x - \bar{x})$$

and since

$$\sum \bar{x}(x - \bar{x}) = \bar{x} \sum (x - \bar{x}) = 0$$

we find that

$${\sum}' xx = \sum x(x - \bar{x}) = \sum x^2 - \bar{x} \sum x = \sum x^2 - \frac{(\sum x)^2}{n} \qquad (2.7)$$

The quantity $(\sum x)^2/n = \bar{x} \sum x = n\bar{x}^2$, using (2.1), is known as the "correction term" and so ${\sum}' xx$ is known as the corrected (for the mean) sum of squares. Combining (2.7) with (2.6) gives computationally convenient forms for s^2 as

$$s^2 = \frac{1}{n-1}\left\{\sum x^2 - \frac{(\sum x)^2}{n}\right\} = \frac{\sum x^2 - \bar{x} \sum x}{n-1} \qquad (2.8)$$

Example: Using the data in the previous example, for which $n = 8$,

(2.4) gives directly

$$s^2 = \frac{1}{7}\{0.0^2 + (-0.3)^2 + 1.4^2 + \cdots + (-0.4)^2\} = \frac{2.46}{7} = 0.35$$

Alternatively, from (2.8)

$$s^2 = \frac{1}{7}\left\{(3.2^2 + 2.9^2 + \cdots + 2.8^2) - \frac{(3.2 + 2.9 + \cdots + 2.8)^2}{8}\right\} = 0.35$$

or, again from (2.8), since $\bar{x} = 3.2$ and $\sum x = 25.6$,

$$s^2 = \frac{1}{7}\{(3.2^2 + 2.9^2 + \cdots + 2.8^2) - (3.2)(25.6)\} = 0.35$$

Sample standard deviation: This is just

$$s = \sqrt{s^2} \qquad (2.9)$$

The sample standard deviation is a measure of sample variability having the same units as the observations themselves.

Coefficient of variation: This is a dimensionless index of variability obtained by expressing the sample standard deviation as a percentage of the sample mean. Thus, when $\bar{x}$ is positive:

$$\text{Coefficient of variation} = 100\left(\frac{s}{\bar{x}}\right)\% \qquad (2.10)$$

$$= 18.5\% \quad \text{for the example above}$$

The coefficient is useful as a descriptive statistic in areas where s may increase with $\bar{x}$.

Range: The range of a sample of observations is the difference between the biggest and the smallest observation ($x_3 - x_4 = x_3 - x_8 = 4.6 - 2.8 = 1.8$ in the above example). For samples of the same size their ranges provide quick, though not the best (unless $n = 2$), measures of sample variability.

§3. SAMPLE AND POPULATION RELATIVE FREQUENCY DISTRIBUTIONS

For observations on a continuous variate, a *sample relative frequency distribution* is an array obtained by:

   (i) conveniently dividing the approximate range of the variate into a number of, usually equal, intervals or classes;

   (ii) showing for each interval its sample relative frequency, that is, the proportion of the sample observations falling into the interval.

Note that it is the variate not the frequency which is distributed.

Histogram: The histogram of a sample frequency distribution is a figure constructed by drawing rectangles, based on each class interval in turn, with *areas* proportional to the respective sample relative frequencies. If, *but only if*, the interval widths are all the same, the heights of the rectangles are, conveniently, proportional to the respective relative frequencies, as in Fig. 3.

Example: The frequency and relative frequency distributions for a sample of 200 goat gestation periods, in days, are given in Table 3.1. There were, for example, 28 goats with gestation periods of between 148.5 and 149.5 days, giving the relative frequency as $28/200 = 0.14$. Figure 3 shows the histogram for the data in Table 3.1.

The relative frequency distribution and the corresponding histogram can suggest interesting characteristics of the parent population. Two common features are: (i) clustering tendency—the clustering of a large proportion of observations in a small interval, for example, near the center of the distribution, and (ii) symmetry—equal proportions of observations in intervals at equal distances from the center. It has been established mathematically that these two features are to be expected

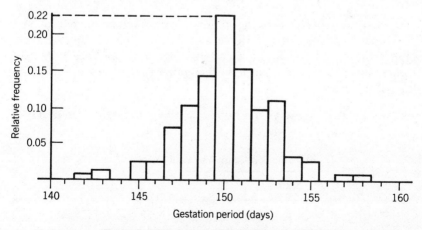

Figure 3. Histogram of 200 goat gestation periods.

Table 3.1. Sample Frequency and Relative Frequency
Distributions for 200 Goat Gestation Periods

Gestation Period (days)	Frequency	Relative Frequency
< 141.5[a]	0	0
142	1	0.005
143	2	0.010
144	0	0
145	4	0.020
146	4	0.020
147	13	0.065
148	20	0.100
149	28	0.140
150	44	0.220
151	31	0.155
152	18	0.090
153	22	0.110
154	6	0.030
155	5	0.025
156	0	0
157	1	0.005
158	1	0.005
> 158.5[a]	0	0
Total	200	1

[a]The symbol < denotes "less than," > denotes "greater than."

when the variability affecting observations is compounded of small components arising from a large number of sources. In such cases observations tend to take on a pattern of great importance in both the theory and practice of statistics. This pattern is that of the normal or Gaussian distribution (§10).

"Stem and leaf" displays, described by their originator Tukey (1977),

Table 3.2. Stem and Leaf Display

Stem	Leaves
2	8899
3	222
4	6

provide an alternative way of exhibiting data characteristics. Table 3.2 shows the simple stem and leaf display for the $n = 8$ observations in the first example.

Table 3.2 shows, for example, the four observations 2.8, 2.8, 2.9, and 2.9 partitioned into the "stem," 2, and the four "leaves," 0.8, 0.8, 0.9, and 0.9, and thus presents the actual values of the observations in each stem class in addition to their number.

Population relative frequency distribution: As previously noted, if a sample of n observations is distributed into classes and n_j is the number of observations in the jth class, the fraction n_j/n is the *relative frequency* for that class. If we erect rectangles on the class intervals, so that the *area* of the jth rectangle is n_j/n, the histogram illustrates the *sample* relative frequency distribution. Also, because $\sum n_j$, the sum of the observations in every class, is equal to n, it follows that $\sum (n_j/n) = 1$; that is, the *total area of the histogram rectangles is unity*.

For bigger and bigger samples we could use progressively smaller class intervals and the histogram rectangles would get thinner and thinner. In the conceptual limit as the sample became infinitely large, growing to become at last equal to the population, the rectangles would become infinitesimally thin and the figure would become a smooth curve showing the *population relative frequency distribution*. This curve retains from the histograms the following two important properties:

(i) If, at any two variate values on the x-axis, we draw ordinate lines to meet the curve, the area enclosed by the two lines, the curve, and the x-axis represents the *population relative frequency for the interval* between the two variate values.

(ii) *The total area under the curve*, that corresponding to the whole range of the variate, *is unity*.

A population relative frequency curve is often defined by a mathematical expression showing how the height of the curve changes according to the value of the variate. Population relative frequency curves are referred to as *probability density functions* because of the relationship between relative frequency and probability.

§4. PROBABILITY (CONTINUOUS VARIATES)

If a single value of a continuous variate is chosen at random from a population, the *probability* that the value lies within any specified interval is the population relative frequency for that interval. If a and b are two specified possible values of a variate x, we write $P[a < x < b]$ for "the probability that a randomly chosen individual x will be greater than a and less than b." *This probability* is *measured by the area under the population relative frequency curve* between the ordinates corresponding to the values a and b as exemplified in §10, Fig. 10.

Good estimates of population parameters are sought therefore because the latter, via their appearance in the mathematical expression defining the probability density function, determine the curve itself and hence the areas measuring probabilities of interest.

More probability concepts, also based on relative frequency ideas and applicable to discrete variate situations, are discussed in §30.

§5. POPULATION MEAN AND VARIANCE: SAMPLE ESTIMATES

Population mean: The inclusion of additional sample members can change the sample mean appreciably when n, the sample size, is small but it is a familiar fact that the sample mean is more stable for large samples. The conceptual limit of the sample mean as the sample is progressively enlarged until it comprises the whole population of possible observations is the *population mean*. Useful notations for the population mean are μ, μ_x, $\mu[x]$, $\mu(x)$, and $E(x)$, the latter being read "the expected value of x."

Population variance: The population variance is one commonly used numerical measure of the variability of members of a population. If x is the variate and μ_x is its population mean, the deviation of x from the population mean is $(x - \mu_x)$ and the square of this deviation is $(x - \mu_x)^2$. The population mean of all these squared deviations is the population variance denoted as σ_x^2 or $V(x)$. Thus,

$$V(x) = \sigma_x^2 = \mu[(x - \mu_x)^2] = E(x - \mu_x)^2 \qquad (5.1)$$

In a high variability situation when there are high proportions of very large and/or very small deviations, there will be a correspondingly high proportion of large values of $(x - \mu_x)^2$. The mean of the $(x - \mu_x)^2$-values, that is, σ_x^2, will accordingly be large. Conversely, in a low variability situation, when most of the x-values are close to their mean μ_x, most of the $(x - \mu_x)^2$-values, and so the variance σ_x^2, will be small.

Population standard deviation: The population standard deviation of a variate x, denoted by σ_x, is simply the square root of the population variance. The standard deviation is therefore an alternative measure of variability with the useful property that it has the same units as those of the x-variate itself.

Sample estimates: The sample mean, the sample variance (2.6), and the sample standard deviation (2.9) are the natural point estimates of the corresponding population quantities. The sample calculations mimic what would be done if it were possible to calculate the population parameters and, in the variance estimate (2.4), we should, if μ_x were known, replace $\bar{x}$ by μ_x and divide $\sum_1^n (x_i - \mu_x)^2$ by n instead of $(n-1)$.

Degrees of freedom (df): When the $x_1, x_2, \ldots, x_n$ are independent, so are the n *deviations*, $(x_1 - \mu_x), \ldots, (x_n - \mu_x)$, from the *population mean*. The n *deviations from the sample mean*, however, are not independent because $\bar{x}$, which is a variate, appears in each deviation and, relatedly, because $\sum_1^n (x - \bar{x}) = 0$, which shows that any one deviation can be obtained as $0 -$ (the sum of the other $n-1$ deviations). It can be shown that $\sum' xx$, the sum of squares of deviations *from* $\bar{x}$, is equivalent, on the average, to the sum of the squares of, not n, but $n-1$ deviations *from* μ_x. This equivalent number of independent deviations is known as the number of degrees of freedom.

It can also be shown that $\sum' xx$ is less than $\sum_1^n (x - \mu_x)^2$ and, again, division by the number of degrees of freedom, rather than by just n, on the average, compensates for the difference to ensure that s^2 is then an *unbiased* estimate of σ^2 in the formal sense that

$$\mu\left[\frac{1}{n-1}\sum (x_i - \bar{x})^2\right] = \mu\left[\frac{1}{n}\sum (x_i - \mu_x)^2\right] = \sigma_x^2 \qquad (5.2)$$

Note: Although s^2 in (2.6) provides an unbiased estimate of σ_x^2, the statistic $s = \sqrt{s^2}$ does not provide an unbiased estimate of the population standard deviation. The bias, however, is usually small enough to support the use of the $\sqrt{s^2}$ estimate in statistical practice.

§6. THE SIMPLE LINEAR TRANSFORMATION OF A VARIATE

Suppose we have a variate x of which the population mean is μ_x and the population variance is σ_x^2. From each x we then calculate a new variate, call it u, related to x by the conversion equation or transformation

$$u = cx + d \qquad (6.1)$$

where c and d are numerical constants. Thus, from a sample $x_1, x_2, \ldots, x_n$ of the x's, we could calculate a sample of u's as $u_1, u_2, \ldots, u_n$, where

$$u_1 = cx_1 + d, \quad u_2 = cx_2 + d, \quad \ldots, \quad u_n = cx_n + d$$

As a practical application we might code data by choosing c and d to give u-values that are computationally more convenient than the x's. The following rule shows how the means, variances, and standard deviations of the u and x variates are related.

RULE 1: If $u = cx + d$, where c and d can be any numerical constants, then

$$\mu_u = c\mu_x + d, \qquad V(cx) = \sigma_u^2 = c^2 V(x), \qquad \sigma_u = |c|\sigma_x \qquad (6.2)$$

where $|c|$ is the positive absolute magnitude of c, that is, $|c| = c$ itself if c is positive, while $|c| = -c$ if c itself is negative.

Similarly, the sample estimates of the respective parameters are connected as

$$\bar{u} = c\bar{x} + d, \qquad s_u^2 = c^2 s_x^2, \qquad s_u = |c|s_x \qquad (6.3)$$

Examples:

(i) If $\mu_x = 10$, $\sigma_x^2 = 5$, and $u = 4x - 3$, that is, $c = 4$ and $d = -3$, then $\mu_u = 4(10) - 3 = 37$, $\sigma_u^2 = (4^2)(5) = 80$, and $\sigma_u = 4(\sqrt{5}) = \sqrt{80}$.

(ii) If $c = 1$ so that $u = x + d$, we have $\mu_u = \mu_x + d$, $\sigma_u^2 = \sigma_x^2$, and $\sigma_u = \sigma_x$.

(iii) Find the mean and variance of the x-sample, 1960, 1910, 1940, and 1990. Subtracting 1900 and dividing by 10 gives the conversion equation $u = (x - 1900)/10 = (1/10)x - 190$. The corresponding u-values are then 6, 1, 4, and 9, giving $\bar{u} = 5$ and $s_u^2 = 34/3$. Using (6.3) in reverse, we then obtain $\bar{x} = (\bar{u} - d)/c = \{5 - (-190)\}/(1/10) = 1950$ and $s_x^2 = s_u^2/c^2 = 3400/3$, with three degrees of freedom.

(iv) A sample variance of Fahrenheit temperature observations was calculated to be 270. Show that, if the observations had been recorded in Celsius units, the sample variance would be 83.3 $(°C)^2$.

Rule 1 shows that if a constant is added to (or subtracted from) each observation, as in Example (ii), the population and sample means are shifted by the same constant, while the population and sample variances and standard deviations are unaltered; if each observation is multiplied by the same constant, the mean is multiplied by the constant, the standard deviation is multiplied by the absolute magnitude of the constant, and the *variance* is multiplied by the *square* of the constant.

§7. THE ADDITION AND SUBTRACTION OF INDEPENDENT VARIATES

So far we have been thinking about a sample each member of which has been drawn from just one population. It is often useful to think of the more general case where x_1 is one random member from a population with mean μ_1 and variance σ_1^2, x_2 is one random member from a population with mean μ_2 and variance σ_2^2, and so on up to, say x_k, one random member from a population with mean μ_k and variance σ_k^2, where the successive means may be, but are not necessarily, the same and similarly for the variances.

For a while independent variates are our main concern and, provided all the x-variates *are* independent, a second simple rule gives the population mean and variance of variates which are linear combinations of these x's as are the variates $(x_1 + x_2)$, $(x_1 - x_2)$ and $c_0 + c_1 x_1 + c_2 x_2 + \cdots + c_k x_k$, where $c_0, c_1, c_2, \ldots, c_k$ represent numerical constants, which can be appropriately chosen for particular practical and theoretical applications.

RULE 2: If x_1 and x_2 are two independently distributed variates with means μ_1 and μ_2 and variances σ_1^2 and σ_2^2, respectively, then

$$\mu(x_1 + x_2) = \mu_1 + \mu_2 \quad \text{and} \quad V(x_1 + x_2) = V(x_1) + V(x_2) = \sigma_1^2 + \sigma_2^2 \quad (7.1)$$

and hence

$$\sigma(x_1 + x_2) = \sqrt{\sigma_1^2 + \sigma_2^2}$$

The variance part of this rule is easily remembered in the form: "The variance of the sum of two independent variates is the sum of the two individual variances."

By applying this rule in conjunction with Rule 1, we can easily find the mean and variance for any linear combination.

Examples: (Assuming all the individual x-variates are independent.)

(i) If $y = c_1 x_1 + c_2 x_2$, then, by Rule 2,

$$\mu_y = \mu(c_1 x_1) + \mu(c_2 x_2) \quad \text{and} \quad V(y) = V(c_1 x_1) + V(c_2 x_2)$$

so that, now using Rule 1,

$$\mu_y = c_1 \mu_1 + c_2 \mu_2 \quad \text{and} \quad V(y) = c_1^2 \sigma_1^2 + c_2^2 \sigma_2^2$$

(ii) If $y = x_1 - x_2$, then from the previous results with $c_1 = 1$ and $c_2 = -1$,

$$\mu_y = \mu_1 - \mu_2 \quad \text{and} \quad V(y) = \sigma_1^2 + \sigma_2^2 = V(x_1 + x_2)$$

so that, when two variates are independent, the population variance of their difference is the same as the population variance of their sum.

(iii) If $y = c_1 x_1 + c_2 x_2 + c_3 x_3$, c_1, c_2, and c_3 being constants, then

$$\mu_y = \mu(c_1 x_1 + c_2 x_2) + \mu(c_3 x_3) = c_1 \mu_1 + c_2 \mu_2 + c_2 \mu_3$$

and

$$V(y) = V(c_1 x_1 + c_2 x_2) + V(c_3 x_3) = c_1^2 \sigma_1^2 + c_2^2 \sigma_2^2 + c_3^2 \sigma_3^2$$

(iv) If $y = 10 + 5 x_1 - 2 x_2 - 4 x_3 + x_4$, then, by repeated applications of Rules 1 and 2 and noting that $V(10) = 0$ because 10 is constant not a variate, we find that

$$\mu_y = 10 + 5 \mu_1 - 2 \mu_2 - 4 \mu_3 + \mu_4$$
$$\sigma_y^2 = V(y) = 25 \sigma_1^2 + 4 \sigma_2^2 + 16 \sigma_3^2 + \sigma_4^2$$

so that,

$$\sigma_y = \sqrt{25 \sigma_1^2 + 4 \sigma_2^2 + 16 \sigma_3^2 + \sigma_4^2}$$

Note: Sample variance results corresponding to the above will appear in later, practical contexts and in §68[+] where analogous results for correlated, that is, not independent, variates are given.

§8. THE DISTRIBUTION OF THE SAMPLE MEAN

Suppose that from some infinite parent population or distribution we take a random sample of n observations and calculate the sample mean, say $\bar{x}_1$. We then take another random sample and find its mean, say $\bar{x}_2$, and so on. We should then be accumulating a second distribution, that of the sample means, the individual members of this derived distribution being $\bar{x}_1, \bar{x}_2, \ldots$. The population mean of this distribution is the same as that of the parent distribution, that is, μ_x. The population variance of the distribution of the sample means, however, is σ_x^2/n and so becomes smaller as the sample size n gets larger. Hence, sample means are less scattered than the original x's and so are more closely clustered around μ_x. Correspondingly, if we take one such sample mean (of two or more observations) it is more likely to be close to the population mean than a single observation is. That is why large samples are recommended for precise estimation.

The results to be remembered are

$$\mu_{\bar{x}} = \mu[\bar{x}] = \mu_x \qquad (8.1)$$

$$\sigma_{\bar{x}}^2 = V(\bar{x}) = \mu[(\bar{x} - \mu_x)^2] = \frac{\sigma_x^2}{n} \qquad (8.2)$$

$$\sigma_{\bar{x}} = \mathrm{sd}(\bar{x}) = \frac{\sigma_x}{\sqrt{n}} \qquad (8.3)$$

These results are easily established using the rules in §6 and §7. First we obtain the related results for sample totals.

Sample total—its population mean and variance: If the sample observations are $x_1, x_2, \ldots, x_n$, we know, from Rule 2 for a sum of independent observations, that

$$\mu[x_1 + x_2 + \cdots + x_n] = \mu_1 + \mu_2 + \cdots + \mu_n$$

$$= n\mu_x \qquad (8.4)$$

because, here, all n of the x's come from just one and the same population, so that $\mu_1 = \mu_2 = \cdots = \mu_x$. Similarly, $V(x_1) = V(x_2) = \cdots = V(x_n) = \sigma_x^2$ and hence, for the variance, Rule 2 gives

$$V\left(\sum x\right) = V(x_1 + x_2 + \cdots + x_n) = V(x_1) + V(x_2) + \cdots + V(x_n)$$

$$= n\sigma_x^2 \qquad (8.5)$$

so that, the larger the sample size n, the more variable are the sample totals.

Sample mean—its population mean and variance: The sample mean $\bar{x}$ and the sample total $\sum x$ are themselves variates; each has a probability density function based on that of the original variate x. And since

$\bar{x} = \sum x/n$, we can apply Rule 1 (multiplication of a variate by a constant), taking the constant $c = 1/n$ here, to get immediately from (8.4) and (8.5)

$$\mu_{\bar{x}} = \left(\frac{1}{n}\right) n\mu_x = \mu_x$$

as in (8.1), and, as in (8.2),

$$V(\bar{x}) = \left(\frac{1}{n}\right)^2 V\left(\sum x\right) = \frac{n\sigma_x^2}{n^2} = \frac{\sigma_x^2}{n}$$

Estimates: Since $s_x^2 = \sum' xx/(n-1)$ in (2.6) is the sample estimate of σ_x^2, the estimation results for the parameters of the distribution of sample means are

$$\bar{x} \text{ is the estimate of } \mu_{\bar{x}} = \mu_x$$

$s_{\bar{x}}^2 = \dfrac{s_x^2}{n}$, with $(n-1)$ degrees of freedom, is the estimate of $\sigma_{\bar{x}}^2$ (8.6)

$$s_{\bar{x}} = \frac{s_x}{\sqrt{n}} \text{ is the estimate of } \sigma_{\bar{x}}$$

$s_{\bar{x}}$ is sometimes termed "the standard error" of the mean.

Examples:

(i) If repeated random samples of size n are taken from a population of x-variate values for which $\mu_x = 17$ and $\sigma_x^2 = 1600$, find the variance of the *sample totals*, $T = \sum_1^n x_i$, for (a) $n = 4$, (b) $n = 9$, and (c) $n = 16$. We find, from (8.5),

(a) $V(T) = \ \ 4(1600) = 16,400$
(b) $V(T) = \ \ 9(1600) = 14,400$
(c) $V(T) = 16(1600) = 25,600$

The bigger n becomes, the more variable are the totals of repeated samples.

(ii) Further to (i) with $\mu_x = 17$ and $\sigma_x^2 = 1600$, find the population mean, variance, and standard deviation of *sample means*, $\bar{x} = T/n$, for (a) $n = 4$, (b) $n = 9$, and (c) $n = 16$.

	(a)	(b)	(c)
(8.1)	$\mu_{\bar{x}} = \mu_x = 17$	$\mu_{\bar{x}} = \mu_x = 17$	$\mu_{\bar{x}} = \mu_x = 17$
(8.2)	$V(\bar{x}) = 1600/4$	$\sigma_{\bar{x}}^2 = 1600/9$	$\sigma_{\bar{x}}^2 = 1600/16$
	$= 400$	$= 177.8$	$= 100$
(8.3)	$\sigma_{\bar{x}} = \sqrt{400}$	$\sigma_{\bar{x}} = \sqrt{1600/9}$	$\sigma_{\bar{x}} = \sqrt{100}$
	$= 20$	$= 13.3$	$= 10$

The bigger n becomes, the more the variability of sample means is reduced. A plot of $\sigma_{\bar{x}}$ against $\sqrt{n}$ nicely illustrates this.

(iii) Suppose that we take three independent samples of size $n = 16$ from the x-population in (i), calculate the sample totals, T_1, T_2, and T_3 and then the contrast-quantity $L = T_1 - (T_2 + T_3)/2$. Suppose we keep repeating this to get a whole sequence of L's: $L_1, L_2, L_3, \ldots$. How variable are these L's? To answer this we can find $V(L)$ as follows: From §7 [compare Example (iii)] taking $T_1 = x_1$, $T_2 = x_2$, and $T_3 = x_3$ as the independent variates,

$$V(L) = V\{T_1 - (T_2/2) - (T_3/2)\} = V(T_1) + V(T_2/2) + V(T_3/2)$$

From (6.2), with $d = 0$ and $c = \frac{1}{2}$,

$$V(T_2/2) = (\tfrac{1}{2})^2 V(T_2) \quad \text{and} \quad V(T_3/2) = (\tfrac{1}{2})^2 V(T_3)$$

so that

$$V(L) = V(T_1) + \tfrac{1}{4}V(T_2) + \tfrac{1}{4}V(T_3)$$

In the numerical case of Example (i) above,

$$V(T_1) = V(T_2) = V(T_3) = n\sigma_x^2 = 16(1600) = 25{,}600$$

Hence,

$$V(L) = (1 + \tfrac{1}{4} + \tfrac{1}{4})(25{,}600) = 38{,}400$$

(iv) Further to (iii), suppose that, for each set of three independent samples, we also calculate the sample means $\bar{x}_1$, $\bar{x}_2$, and $\bar{x}_3$ and the contrast $l = \bar{x}_1 - (\bar{x}_2 + \bar{x}_3)/2$. Repetitions will then generate the sequence $l_1, l_2, l_3, \ldots$. What is the variance of the conceptual population of these l's? Again, as in (iii), we proceed as follows: from §7, because $\bar{x}_1$, $\bar{x}_2$, and $\bar{x}_3$ are independent,

$$V(l) = V\{\bar{x}_1 - (\bar{x}_2/2) - (\bar{x}_3/2)\} = V(\bar{x}_1) + V(\bar{x}_2/2) + V(\bar{x}_3/2)$$

From (6.2) with $d = 0$ and $c = \frac{1}{2}$,

$$V(\bar{x}_2/2) = (\tfrac{1}{2})^2 V(\bar{x}_2) \quad \text{and} \quad (V(\bar{x}_3/2) = (\tfrac{1}{2})^2 V(\bar{x}_3)$$

For means of samples of size n therefore

$$V(l) = \left(1 + \frac{1}{4} + \frac{1}{4}\right)\left(\frac{\sigma_x^2}{n}\right)$$

And, in the particular case of Example (ii),

$$V(l) = \left(1 + \frac{1}{4} + \frac{1}{4}\right)\left(\frac{1600}{16}\right) = 150$$

Note: Because $l = L/16$, $V(l)$ can be directly derived from $V(L)$, and conversely, using (6.2). Thus, $V(l) = (\tfrac{1}{16})^2 V(L) = 38{,}400/256 = 150$.

(v) If a variate u is simply a constant multiplier of another variate x, with the conversion equation $u = cx$, where c is a numerical constant, then

$$\frac{u^2}{\sigma_u^2} = \frac{(cx)^2}{c^2 \sigma_x^2} = \frac{x^2}{\sigma_x^2}$$

and, taking square roots,

$$\frac{u}{\sigma_u} = \frac{x}{\sigma_x} \tag{8.7}$$

Thus, in the preceding example, where $l = \frac{1}{16}L$, taking $u = l$, $x = L$, and $c = \frac{1}{16}$ shows that

$$\frac{l}{\sigma_l} = \frac{L}{\sigma_L}$$

The corresponding result, using estimates of the standard deviations, will be found useful in later data analyses.

§9. STATISTICAL MODELS

Figure 9 illustrates that, if x_i is an observation from a distribution with population mean μ_x, the distance x_i from the origin 0 is

$$x_i = \mu_x + (x_i - \mu_x)$$

where $(x_i - \mu_x)$, the deviation of x_i from its population mean, is positive if $x_i > \mu_x$ and negative if $x_i < \mu_x$. Writing $\varepsilon_i = x_i - \mu_x$, we then have

$$x_i = \mu_x + \varepsilon_i \tag{9.1}$$

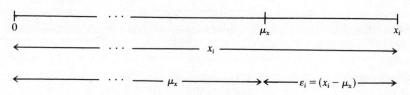

Figure 9. The statistical model of an observation.

which is the simple statistical model for the observation x_i. The equation shows that each x_i is made up of two added components, the population mean μ_x, which will be the same for all observations, and the individual deviation ε_i, which will differ from one observation to another.

In fact the ε_i's are themselves variates and for their distribution we find by §6, Rule 1,

$$\mu(\varepsilon) = \mu[x - \mu_x] = \mu_x - \mu_x = 0 \qquad (9.2)$$

and, since the subtraction of the constant μ_x does not change variance,

$$\sigma_\varepsilon^2 = V(\varepsilon) = V(x) = \sigma_x^2 \qquad (9.3)$$

The statistical model for a sample mean is obtained, using (9.1), as

$$\bar{x} = \frac{\sum_1^n (\mu_x + \varepsilon_i)}{n} = \mu_x + \bar{\varepsilon} \qquad (9.4)$$

where

$$\mu(\bar{\varepsilon}) = 0 \quad \text{and} \quad V(\bar{\varepsilon}) = \frac{\sigma_x^2}{n} \qquad (9.5)$$

More complicated statistical models are built up to represent other situations where statistical analyses are used. Generally, the models are idealized, rather than exact, representations of phenomena and much of the art consists in devising models that are tractable to analysis and sufficiently close as representations of the phenomena under study.

§10. NORMAL (GAUSSIAN) DISTRIBUTIONS

Characteristic contributions of statistical methods to research turn on the attachment of moderating probabilities to inferences from sample observations. For example, it is common that a probability-based assessment is required for a statement that a given observed difference between two sample means is attributable merely to chance or is a signal of a real difference between the two population means. Later, we shall see how to calculate confidence intervals for population means of interest. The widths of such intervals depend on the sample size, the variability, and the prespecified confidence level. Such probability statements are commonly based on the assumption that the parent distribution, of which the observations constitute a random sample, is one of several standard or reference distributions. Probabilities in many practical applications of statistics are obtained from *normal* parent distributions and reference distributions derived from them.

The distinction between the *variate population* and the *variate distribution* is worth noting. The population is the simple aggregate of all the variates not arranged in any special way; the distribution consists of the same variates arranged by size from the smallest to the largest so that, knowing the distribution, it is in principle possible to ascertain the proportion of the variates lying between any specified limits.

For a continuous variate, the distribution is mathematically specified by the probability density function or, equivalently, by the cumulative distribution function.

Probability density function: In the statement "the probability that a random value of the variate falls in the interval from x to $(x + dx)$ is $f(x)dx$," $f(x)$ is the probability density at the value x and dx formally represents the width of an infinitesimally narrow interval.

Cumulative distribution function: In the statement "the probability that a random value of the variate is less than the value a is $F(a)$," $F(a)$ is the value which the cumulative distribution function takes at the point $x = a$; $F(a)$ measures the area under the probability density curve to the left of the specified value a. $F(a)$ is obtained as the sum or integration of the incremental $f(x)dx$ probability contributions for all the variate values from $x = -\infty$ up to $x = a$.

Normal distribution probability density function: If x is a normally distributed variate, the expression for its probability density function is

$$f(x) = \text{probability density at the point } x = \frac{1}{\sqrt{2\pi\sigma_x^2}}\, e^{-(x-\mu_x)^2/2\sigma_x^2} \quad (10.1)$$

wherein the constant $e = 2.718 \cdots$ is the base of natural logarithms. Thus, if μ_x and σ_x^2, the population mean and variance, were known, we could find $f(x)$ for any given numerical value of x and, by plotting

$y = f(x)$ against x, obtain a curve which would have the bell-section form of Fig. 10.

The curve is highest when $x = \mu_x$ which, because of the symmetry, is the mean of the distribution. For x-values smaller or larger than μ_x, the height of the curve diminishes symmetrically and tends progressively towards zero as x-values become either extremely large or extremely small.

The shaded area between $x = a$ and $x = b$ in Fig. 10 represents the probability that a randomly obtained value of the x-variate will lie between the two values a and b. As shown, this is written $P(a < x < b)$ and if $F(x)$ is the normal cumulative distribution function we see from its definition above that

$$P(a < x < b) = F(b) - F(a) \qquad (10.2)$$

It can be seen from Fig. 10 that, because the curve is high for x-values close to μ_x, the probabilities of getting such x-values by random sampling are relatively high and that the probabilities tail off to almost zero for regions of extremely large positive or negative deviations. The practical force of this when μ_x is unknown is simply that if an observation x is close to its mean μ_x then μ_x is close to the known x-observation. Hence, for a good estimation of μ_x, the higher the probability of small $|x - \mu|$-values or $(x - \mu_x)^2$-values, the better.

Of special importance is the fact that if each of a number of variates is normally distributed then linear combinations of variates, such as those discussed in §7, are also normally distributed. The sample mean $\bar{x} = (x_1 + x_2 + \cdots + x_n)/n$ of n random observations is a simple linear combination of the observations $x_1, x_2, \ldots, x_n$. Hence, if each of the x's is

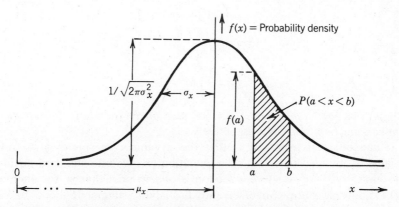

Figure 10. A normal distribution probability density curve.

normally distributed, with mean μ_x and variance σ_x^2, the sample mean is also normally distributed and, from §8, the population mean and variance of such sample means are μ_x and σ_x^2/n, respectively.

A useful notational convention is to write

$$x \sim N(\mu_x, \sigma_x^2)$$

for "the variate x is distributed normally with population mean μ_x and population variance σ_x^2." Correspondingly, if $\bar{x}$ is the mean of a sample of n random, independent, normally distributed x-values, we write $\bar{x} \sim N(\mu_x, \sigma_x^2/n)$. And, to specify a random sample of n normally distributed observations, we write

$$x_i \sim NI(\mu_x, \sigma_x^2) \qquad i = 1, \ldots, n$$

introducing the I to specify the independence of the observations.

§11. THE STANDARD NORMAL (GAUSSIAN) REFERENCE DISTRIBUTION

If $x \sim N(\mu, \sigma^2)$ and from each value x a corresponding variate z is obtained by the conversion equation

$$z = \frac{x - \mu}{\sigma} = \left(\frac{1}{\sigma}\right)x - \frac{\mu}{\sigma} \qquad (11.1)$$

the normally distributed variate z is known as a "standard normal variate" and its distribution is the "standard normal distribution." From (6.2) with $u = z$, $c = 1/\sigma$, and $d = -\mu/\sigma$, it follows that

$$\mu_z = \left(\frac{1}{\sigma}\right)\mu - \frac{\mu}{\sigma} = 0 \qquad (11.2)$$

and

$$\sigma_z^2 = \left(\frac{1}{\sigma}\right)^2 \sigma^2 = 1 \qquad (11.3)$$

that is, *the population mean and variance of the standard normal distribution are zero and one*, respectively.

It can be seen from (11.1) that the z-value corresponding to any x-value is obtained by scaling the deviation $(x - \mu)$ into standard deviation units. And, since the conversion is one to one, that is, for each x there is but one z and conversely, any probability statement about values of the x-distribution has an exact equivalent for corresponding values of the z-distribution. For example, if z_1 and z_2 are the correspondents to x_1 and x_2, so that,

$$z_1 = \frac{x_1 - \mu}{\sigma} \quad \text{and} \quad z_2 = \frac{x_2 - \mu}{\sigma}$$

then every x between x_1 and x_2 has a z-correspondent between z_1 and z_2 to give the equal probabilities

$$P[x_1 < x < x_2] = P[z_1 < z < z_2] \tag{11.4}$$

And equivalently, since $x_1 = \mu + \sigma z_1$ if $z_1 = (x_1 - \mu)/\sigma$, and similarly $x_2 = \mu + \sigma z_2$,

$$P[z_1 < z < z_2] = P[\mu + \sigma z_1 < x < \mu + \sigma z_2] \tag{11.5}$$

For example, it is known that if $z_1 = -1.96$ and $z_2 = +1.96$,

$$P[-1.96 < z < +1.96] = 0.95$$

that is, 95% of all z-variates lie between -1.96 and $+1.96$. Hence, if $x \sim N(\mu, \sigma^2)$, putting $z_1 = -1.96$ and $z_2 = +1.96$ in (11.5) shows that

$$P[\mu - 1.96\sigma < x < \mu + 1.96\sigma] = 0.95 \tag{11.6}$$

This shows that 95% of the normally distributed x-values fall in an interval of width $2(1.96)\sigma$ symmetrically disposed about μ. Consistently with the use of σ and σ^2 as measures of variability, this shows that the 95% containing intervals get wider as σ increases.

The table of z-variate probabilities: Reference probability tables are used in practical applications of statistics to find probabilities corresponding to specified variate values and variate values corresponding to specified probabilities. Figure 11 illustrates the z-variate and probability correspondence used in Table A2.

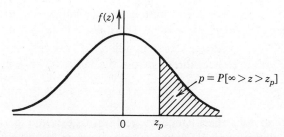

Figure 11. Illustration of the standard normal variate tabulation.

The Fig. 11 probability density curve shows the two quantities tabulated: (i) the right-hand tail area of the distribution, p, which is the probability that a randomly chosen member of the z-distribution will exceed, that is, fall to the right of (ii), the abscissa value z_p in Fig. 11. The probabilities p are given in the body of the table corresponding to the variate values z_p in the left column, for the first decimal place, and the top row, for the second decimal place. Thus, taking z_p to be 1.96, we read in the body of the table that $p = 0.025$, giving the probability statement $P[z > 1.96] = 0.025$. Related probabilities are immediately obtainable using the facts that (a) normal distributions are symmetrical and (b) the total area under the curve is unity.

Hence, using (a) we can immediately obtain the left-tail probability that $P[z < -1.96] = 0.025$ also and, generally,

$$P[-\infty < z < -z_p] = P[\infty > z > z_p] = p \qquad (11.7)$$

Using (b) we note that

$$P[z < z_p] = 1 - P[z > z_p] = 1 - p \qquad (11.8)$$

and also that, since $P[z > -z_p] = 1 - p$,

$$z_{1-p} = -z_p \qquad (11.9)$$

Accordingly, $P[z < 1.96] = 1 - 0.025 = 0.975$ and $z_{0.975} = -1.96$. Using (a) and (b) we see that $P[z < 0] = P[z > 0] = \frac{1}{2}$ so that

$$P[0 < z < z_p] = \frac{1}{2} - P[z > z_p] = 0.5 - p \qquad (11.10)$$

and, in particular, $P[0 < z < 1.96] = 0.5 - 0.025 = 0.475$.

Note: The development of accurate table reading is facilitated by drawing sketches showing the probabilities required.

Examples:

(i) Suppose that goat gestation periods x are normally distributed with $\mu = 150$ days and $\sigma = 3$ days. Find the probabilities that a randomly selected gestation period will be (a) greater than 154 days, (b) between 144 and 154 days, and (c) less than 144 days.

(a) The z-variate corresponding to $x = 154$ is $(154 - 150)/3 = 1.33$. Hence, directly from the table,

$$P[x > 154] = P[z > 1.33] = 0.092$$

so that the chance of getting a gestation period as long or longer than 154 days is less than 1 in 10.

(b) The z-variate corresponding to 144 days is $(144 - 150)/3 = -2$. We can then proceed to obtain

$$P[144 < x < 154] = P[-2 < z < 1.33]$$

Then, by the symmetry property,

$$P[-2 < z < 0] = P[0 < z < 2] = 0.5 - P(z > 2)$$
$$= 0.5 - 0.023 = 0.477$$

and, similarly,

$$P[0 < z < 1.33] = 0.5 - 0.092 = 0.408$$

so that, adding the two probabilities,

$$P[144 < x < 154] = 0.477 + 0.408 = 0.885$$

(c) For this we simply write

$$P[x < 144] = P[z < -2] = P[z > 2] = 0.023$$

(ii) If $x \sim N(\mu, 9)$, find the distance d for which we can say the probability is 0.95 that the mean of a random sample of $n = 25$ observations will be no further from μ, in either direction, than d. Since $\bar{x} \sim N(\mu, 9/25)$, from (8.2), is now the variate of interest, $(\bar{x} - \mu)$-deviations must now be scaled by

$$\sigma_{\bar{x}} = \sqrt{\frac{\sigma^2}{n}} = \sqrt{\frac{9}{25}} = 0.6$$

to get z-values corresponding to $\bar{x}$-values. We need to find d such that

$$P[\mu - d < \bar{x} < \mu + d] = 0.95$$

The z-value corresponding to $\mu - d$ is

$$z_1 = \frac{(\mu - d) - \mu}{\sigma_{\bar{x}}} = \frac{-d}{\sigma_{\bar{x}}} = \frac{-d}{0.6}$$

and, similarly, z_2 corresponding to $(\mu + d)$ is $d/0.6$. We therefore need

$$P[-d/0.6 < z < d/0.6] = P[z_1 < z < z_2] = 0.95$$

It is now found from the z-table that $z_1 = -1.96$ and $z_2 = 1.96$ so that $d/0.6 = 1.96$, giving $d = 1.18$.

§12. *t*-DISTRIBUTIONS

Applications of standard normal distribution probabilities to practical situations are limited because μ and σ^2 are usually unknown. Many practical problems, however, can be dealt with by reference to *t*-distributions, which are similar to the standard normal distribution and for which probability tables are also available.

A *t*-variate may be recognized from its ratio form $(A - B)/C$, where A is a normally distributed variate, B is the population mean of the variate A, C^2 is the *estimate* of the variance of A so that C itself is the standard deviation.

t-**Variate:** If $y \sim N(\mu_y, \sigma_y^2)$ and s_y^2, with f degrees of freedom, is independent of y and is an unbiased estimate of σ_y^2, then the quantity

$$t_f = \frac{y - \mu_y}{s_y} \tag{12.1}$$

is known as a *t*-variate; it is a member of that *t*-distribution with f degrees of freedom.

The difference from standard normal variates is that deviations from the population mean are not now divided by σ, the population standard deviation, but by s the *estimate* (2.9) of σ. The higher the number of degrees of freedom the closer s is to σ and, in fact, if $f = \infty$, $s = \sigma$ and the *t*-distribution with $f = \infty$ degrees of freedom is exactly the standard normal distribution. The number of degrees of freedom, f, is determined by the way in which the estimate of variance, s_y^2, is calculated. Thus, $f = n - 1$ if $s_y^2 = \sum (y - \bar{y})^2/(n - 1)$; other values for f will occur later.

Because normal $(y - \mu_y)$-deviates are symmetrically distributed about zero, so are *t*-variates; *t*-distributions peak less sharply at the origin, however, and are thicker in the tails than is the *z*-distribution. Hence, $t_p > z_p$ for the same right-tail probability p.

Reading *t*-tables: There are many *t*-distributions, one for each number of degrees of freedom $f = 1, 2, \ldots$. In Table A3, the number of degrees of freedom for the variance estimate involved is located in the first column and, along the corresponding row, values t_p are found beneath right-tail probabilities p, given in the heading row. Thus, if s^2 is calculated from 10 observations, the $f = 9$ row of the table gives $t_{0.05} = 1.833$ for $p = 0.05$ so that the probability is 0.05 that a random *t*-variate with nine degrees of freedom will exceed 1.833. Formally, to display the number of degrees of freedom, we can write $t(f; p)$ for t_p and the

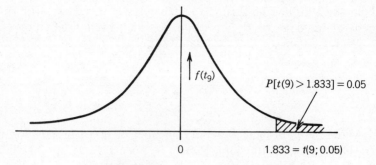

Figure 12. Illustration of t-variate tabulation.

tabulated values are such that

$$P[t > t(f; p)] = P[t > t_p] = p \qquad (12.2)$$

as shown in Fig. 12.

Again, related probabilities are directly obtainable; thus, the total area = unity property gives $P[t < t(9; 0.05)] = 1 - 0.05 = 0.95$. And, two-tail probabilities are available using the symmetry, so that

$$P[|t| > t_p] = P[t < -t_p] + P[t > t_p]$$
$$= 2P[t > t_p] = 2p \qquad (12.3)$$

Equivalently, because the total probability is unity,

$$P[-t_p < t < t_p] = 1 - 2p \qquad (12.4)$$

Thus, because for nine degrees of freedom, $t_{0.025} = 2.262$,

$$P[|t| > 2.262] = P[t < -2.262 \text{ or } t > 2.262] = 2(0.025) = 0.05$$

and

$$P[-2.262 < t < 2.262] = 0.95$$

so that 95% of t-variates with nine degrees of freedom lie between -2.262 and $+2.262$.

§13. CONFIDENCE INTERVALS FOR THE POPULATION MEAN OF A NORMAL DISTRIBUTION AND A GENERAL FORMULATION

The notation $x_i \sim NI(\mu_x, \sigma_x^2)$, $i = 1, 2, \ldots, n$, indicates that the n values $x_1, x_2, \ldots, x_n$ constitute a random sample of normally and independently distributed observations for which μ_x and σ_x^2 are the population mean and variance. From such a sample, estimates $\bar{x}$ of μ_x and s_x^2, with $(n-1)$ degrees of freedom, of σ_x^2 are calculated and, by the definition (12.1) with $y = \bar{x}$, we then see that the quantity $(\bar{x} - \mu_{\bar{x}})/s_{\bar{x}}$ is a t-variate. Hence, because $\mu_{\bar{x}} = \mu_x$ [Eq. (8.1)] and because $s_{\bar{x}}^2 = s_x^2/n$ [Eq. (8.6)] is the estimate of $\sigma_{\bar{x}}^2$,

$$\frac{\bar{x} - \mu_x}{s_{\bar{x}}} = \frac{\bar{x} - \mu_x}{\sqrt{s_x^2/n}} \tag{13.1}$$

is a t-variate and, in fact, a random member of that t-distribution with $f = (n-1)$ degrees of freedom. Hence, if, for a specified probability α, $t_{\alpha/2} = t(n-1; \alpha/2)$ is read from Table A3 so that $P[t > t_{\alpha/2}] = \alpha/2$, we have from (12.4), with $p = \alpha/2$,

$$P[-t_{\alpha/2} < (\bar{x} - \mu_x)/s_{\bar{x}} < t_{\alpha/2}] = 1 - \alpha \tag{13.2}$$

that is,

$$P[-t_{\alpha/2}s_{\bar{x}} < \bar{x} - \mu_x < t_{\alpha/2}s_{\bar{x}}] = 1 - \alpha$$

and since adding μ_x to each element on the left will not change the probability,

$$P[\mu_x - t_{\alpha/2}s_{\bar{x}} < \bar{x} < \mu_x + t_{\alpha/2}s_{\bar{x}}] = 1 - \alpha$$

Now,

$$\mu_x - t_{\alpha/2}s_{\bar{x}} < \bar{x} \quad \text{implies that} \quad \mu_x < \bar{x} + t_{\alpha/2}s_{\bar{x}}$$

and

$$\bar{x} < \mu_x + t_{\alpha/2}s_{\bar{x}} \quad \text{implies that} \quad \bar{x} - t_{\alpha/2}s_{\bar{x}} < \mu_x$$

so that a probability statement entirely equivalent to (13.2) is that

$$P[\bar{x} - t_{\alpha/2}s_{\bar{x}} < \mu_x < \bar{x} + t_{\alpha/2}s_{\bar{x}}] = 1 - \alpha \tag{13.3}$$

This leads to the following definition:

The $100(1 - \alpha)\%$ confidence interval for the mean of a normal distribution is that from $\bar{x} - t_{\alpha/2}s_{\bar{x}}$ to $\bar{x} + t_{\alpha/2}s_{\bar{x}}$, where $t_{\alpha/2} = t(n-1; \alpha/2)$ and $s_{\bar{x}} = \sqrt{s_x^2/n}$. The interval may also be written

$$\bar{x} \mp t_{\alpha/2}\sqrt{\frac{s_x^2}{n}} \tag{13.4}$$

the negative and positive signs being taken to obtain the lower and upper limits of the interval, respectively.

The correct interpretation of (13.3) and the interval (13.4) is that if an infinite number of random samples of size n were to be taken and confidence intervals were worked out each time, then $100(1 - \alpha)\%$ of the intervals would contain μ_x and $100\alpha\%$ of the intervals would fail to bracket μ_x. Hence, *before one particular random sample is taken*, the probability that the interval to be realized will contain μ_x is $(1 - \alpha)$. *After the interval has been calculated* it does or does not contain μ_x: we do not know. However, we can say that μ_x *is* in the interval unless an event with probability less than α occurred.

The width of the $100(1 - \alpha)\%$ confidence interval is $2t_{\alpha/2}\sqrt{s_x^2/n}$; it increases with the variability measure s_x^2, decreases as n is increased, and increases, via $t_{\alpha/2}$, if more confidence is required.

Note: If calculated confidence intervals include a range of values that are known to be impossible, such as negative values for a necessarily positive μ_x, the interval is contracted to exclude the impossible values.

Confidence intervals—a general formulation: Suppose that a population parameter, call it θ, is estimated unbiasedly by a quantity $\hat{\theta}$ which is a linear combination of normally distributed individual observations. Suppose also that we can obtain the estimated variance of $\hat{\theta}$ as $\hat{V}(\hat{\theta}) = s_{\hat{\theta}}^2 = cs^2$, where c is a constant that depends on the particular structure of $\hat{\theta}$, and s^2, with f degrees of freedom, is the unbiased estimate of the variance of an individual observation. Then $(\hat{\theta} - \theta)/\sqrt{cs^2}$ is a $t(f)$-variate and, by exactly the same argument as that above, **the $100(1 - \alpha)\%$ confidence interval for θ is**

$$\hat{\theta} \mp t(f; \alpha/2)s_{\hat{\theta}} = \hat{\theta} \mp t(f; \alpha/2)\sqrt{cs^2} \qquad (13.5)$$

that is, literally,

$$\text{(the point estimate)} \mp t_{\alpha/2} \text{ (the standard deviation estimate}$$

$$\text{for the point estimate)} \qquad (13.6)$$

Result (13.4) is one example with $\theta = \mu_x$, $\hat{\theta} = \bar{x}$, $c = 1/n$, $s_{\bar{x}} = \sqrt{s^2/n}$, and $f = (n - 1)$; many other examples will occur subsequently.

Example: If $x_i \sim NI(\mu, \sigma^2)$, $i = 1, \ldots, n = 10$, gave $\bar{x} = 12.5$ and $s^2 = 160$ (nine degrees of freedom), calculate a 95% confidence interval for μ.

Here $f = 9$ and $\alpha = 0.05$, if $1 - \alpha = 0.95$, giving $t_{\alpha/2} = t(9; 0.025) = 2.262$. The $100(1 - 0.05) = 95\%$ confidence interval is then, from (13.4),

$$12.5 \mp 2.262 \sqrt{\frac{160}{10}} = 3.45, 21.55$$

After rounding to avoid illusory precision, we can infer that unless we have been afflicted by a less than 1 in 20 chance event, μ lies somewhere between 3 and 22.

Note: Confidence interval limits have the same units as the observations.

§14. CHI-SQUARED (χ^2) REFERENCE DISTRIBUTIONS

To obtain confidence intervals for the population variance parameter of a normal distribution, probabilities from another, the chi-squared, set of reference distributions are required.

The square of a standard normal deviate z is a $\chi^2(1)$-variate; it is a member of that χ^2-distribution with one degree of freedom. Conceptually, if every member of the z-distribution were to be squared, these squares would comprise the χ_1^2 distribution. Thus, if $x \sim N(\mu, \sigma^2)$, where "$\sim$" denotes "is distributed as" or "has the distributional pattern of,"

$$\left(\frac{x-\mu}{\sigma}\right)^2 = \frac{(x-\mu)^2}{\sigma^2} = z^2 \sim \chi_1^2 \tag{14.1}$$

Because of the squaring, all $\chi^2(1)$-variates are positive; they lie between 0 and ∞. Also, the population mean of the $\chi^2(1)$ distribution is 1 because

$$\mu(z) = 0 \quad \text{from (11.2) and hence} \quad \mu(z^2) = V(z) = 1$$

from (11.3) and the definition of variance (5.1).

If repeated random samples of size two are taken from a $\chi^2(1)$-population, each sample total is a member of the $\chi^2(2)$-population: that chi-squared distribution with two degrees of freedom. Equivalently, if z_1 and z_2 are two random standard normal variates, we can write

$$z_1^2 + z_2^2 \sim \chi^2(2)$$

By an easily envisaged extension, the distribution of the sum of f random and independent $\chi^2(1)$-variates is the chi-squared distribution with f degrees of freedom and the population mean of this distribution is $\sum_1^f (1) = f$. We can write

$$z_1^2 + z_2^2 + \cdots + z_f^2 \sim \chi^2(f) \tag{14.2}$$

Reading χ^2-tables: In Table A4, the number of degrees of freedom f is located in the first column and, along the corresponding row, values $\chi^2(f; p) = \chi_p^2$ are found beneath the right-tail probabilities p, given in the heading row. Thus, for $f = 1$ we find $P[\chi^2(1) > 3.841] = 0.05$. This

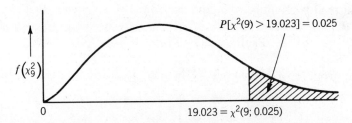

$$P[\chi^2(9) > 19.023] = 0.025$$

$f\left(\chi^2_9\right)$

$$0 \qquad\qquad\qquad 19.023 = \chi^2(9; 0.025)$$

Figure 14. Illustration of χ^2-variate tabulation.

is consistent with (14.1) because $P[z < -1.96$ or $z > 1.96] = 2P[z > 1.96] = 0.05$ and if either $z < -1.96$ or $z > 1.96$, then $z^2 > 1.96^2 = 3.841$. For $f = 9$ and $p = 0.025$ we find $P[\chi^2 > \chi^2_{0.025} = \chi^2(9; 0.025) = 19.023] = 0.025$ as shown in Fig. 14. Only 2.5% of random $\chi^2(9)$-variates exceed 19.023.

An important fact about independent chi-squared variates is their additivity property: if $\chi^2(f_1)$ and $\chi^2(f_2)$ are independent random members of the chi-squared distributions with f_1 and f_2 degrees of freedom, their sum is also a chi-squared variate; in fact, as the sum of $f_1 + f_2$ independent z^2's, it is a $\chi^2(f_1 + f_2)$-variate. Thus,

$$\chi^2(f_1) + \chi^2(f_2) \sim z_1^2 + z_2^2 + \cdots + z_{f_1+f_2}^2 \sim \chi^2(f_1 + f_2) \qquad (14.3)$$

For example, the probability is 0.025 that the sum of a randomly chosen $\chi^2(4)$-variate and a randomly chosen $\chi^2(5)$-variate will exceed 19.023.

§15. CONFIDENCE INTERVALS FOR THE POPULATION VARIANCE OF A NORMAL DISTRIBUTION

A previously noted theoretical result (§5, Degrees of freedom) is that, if $x_i \sim NI(\mu, \sigma^2)$, $i = 1, \ldots, n$, the distribution of the sum of squared deviations, $\sum' xx = \sum_1^n (x_i - \bar{x})^2$, is equivalent to the distribution of a sum of $(n - 1)$ independent $(x - \mu)^2$-values. Hence, dividing both by σ^2,

$$\frac{1}{\sigma^2} \sum_1^n (x_i - \bar{x})^2 \sim \frac{1}{\sigma^2} \sum_1^{n-1} (x_j - \mu)^2 = \sum_1^{n-1} \left(\frac{x_j - \mu}{\sigma}\right)^2$$

and this, as the sum of $(n - 1)$ z^2-values, has the χ^2-distribution with $(n - 1)$ degrees of freedom by the definition (14.2). Beginning then with the fact that

$$\frac{\sum' xx}{\sigma^2} \text{ is a random member of the } \chi^2(n - 1)\text{-distribution} \qquad (15.1)$$

we can read two numbers, a_1 and a_2, from Table A4 such that

$$P[\chi^2_{n-1} < a_1] = P[\chi^2_{n-1} > a_2] = \frac{\alpha}{2}$$

so that, from (15.1) with $\sum' xx = \sum (x - \bar{x})^2$,

$$P\left[a_1 < \frac{\sum' xx}{\sigma^2} < a_2\right] = 1 - \frac{\alpha}{2} - \frac{\alpha}{2} = 1 - \alpha \qquad (15.2)$$

Now, if $a_1 < \sum' xx/\sigma^2 < a_2$, then $\sigma^2 < \sum' xx/a_1$ and $\sum' xx/a_2 < \sigma^2$ so that a probability statement exactly equivalent to (15.2) is that

$$P\left[\frac{\sum' xx}{a_2} < \sigma^2 < \frac{\sum' xx}{a_1}\right] = 1 - \alpha \qquad (15.3)$$

The interpretation is like that in §13. Therefore, unless an event with probability less than α occurs to give a misleading unlikely value for $\sum' xx$ from our sample, σ^2 will lie between the limits $\sum' xx/a_2$ and $\sum' xx/a_1$.

Conventionally, a_1 and a_2 are defined according to the right-hand tail areas as $a_1 = \chi^2_{1-(\alpha/2)}$, so that $P[\chi^2 > \chi^2_{1-(\alpha/2)}] = 1 - (\alpha/2)$, and $a_2 = \chi^2_{\alpha/2}$, so that $P[\chi^2 > \chi^2_{\alpha/2}] = \alpha/2$. We then have that the

$100(1 - \alpha)$% confidence interval for the population variance is from

$$\frac{\sum' xx}{\chi^2_{\alpha/2}} \text{ to } \frac{\sum' xx}{\chi^2_{1-(\alpha/2)}} \qquad (15.4)$$

where the denominators are read from the $\chi^2(n-1)$-tabulation in Table A4.

Example: From a random sample of $n = 11$ normally distributed observations it was calculated that $s^2 = 4$. Find a 95% confidence interval for σ^2_x.

Since $s^2 = 4$, $\sum' xx = (n-1)s^2 = 40$. Also, for a 95% confidence interval, $\alpha/2 = 0.025$, and from the χ^2_{10} table we read $\chi^2_{0.025} = 20.483$ and $\chi^2_{0.975} = 3.247$. From (15.4) the interval is then from $40/20.483 = 1.95$ to $40/3.247 = 12.32$. We say that $1.95 < \sigma^2 < 12.32$ unless an event with probability less than 0.05 has occurred.

Notes:

(i) The intervals are wide when s^2 is large, especially if the number of degrees of freedom is small.

(ii) s^2 does not lie at the interval center.

(iii) For a refined procedure, which uses special tables of alternative divisors of $\sum' xx$ to give generally narrower intervals, see Tate and Klett (1959).

§16. HYPOTHESIS TESTING (SIGNIFICANCE TESTING)

It is often required to test whether or not the evidence in a random sample of observations supports or conflicts with some hypothesis or theory about the unknown value of a parameter. Thus, we might wish to test if $\bar{x} = 4.8$ is statistically compatible with a hypothesis that $\mu_x = 5$. The particular hypothesis to be tested is referred to as the *test hypothesis* H_T or the *null hypothesis* H_0.

Test statistic: Working provisionally on the assumption that H_T is true, a *test statistic* is calculated from the data as an index or measure which is *sensitive to discrepancy from* H_T. Extreme values of the test statistic are unlikely if H_T is true and lead to its rejection as a statement about the real situation being studied.

The *P*-value or the actual significance level or the exceedance probability: This (cogently discussed by Gibbons and Pratt, 1975) is the probability, *determined on the presumption that H_T is true*, of randomly obtaining a test statistic value more extreme than that calculated from the sample. If this P-value is very small we naturally prefer another hypothesis termed the *alternative hypothesis H_A*, and it is formally convenient to say that "H_T is rejected and H_A is accepted." Conversely, if, presuming H_T, the P-value for the test statistic is high, it is better to stay with the test hypothesis and reject H_A.

Type I error: Statistical procedures cannot *prove* that a hypothesis is true or false. If, when H_T is actually true, the accidents of random sampling give a very extreme test statistic value, we shall be deceived by the associated low P-value into making a *Type I error*: *erroneously rejecting a true test hypothesis*.

The probability of making a Type I error, termed the *significance criterion α*, is entirely optional; α is the maximum P-value at which *we* decide, in advance, to reject H_T; α-values of 0.05 and 0.01 are often used.

Critical value, critical region: The critical value(s) is that (are those) for which the P-value is just equal to the rejection significance criterion α.

More extreme test statistic values, those having *P*-values even less than α, are said to lie in the *critical region*. Accordingly, H_T *is rejected if the test statistic falls in the critical region.*

The steps in hypothesis testing can be formalized as follows:

1. Define the hypothesis specification to be examined, by specifying the test and alternative hypotheses and α, the size (probability) of the Type I error to be tolerated.
2. Define and calculate the test statistic, the index of discrepancy from H_T.
3. State the assumption made about the parent distribution of the data. This determines the distribution of the test statistic on the assumption that the test hypothesis is true. This in turn determines the critical region.
4. Define the critical region which corresponds to the α level specified in step 1. Check to see whether or not the test statistic falls in the critical region and reject or accept the test hypothesis accordingly.
5. Find and report the *P*-value.

Notes: Significance tests are advisedly guides, not the final arbiters, in practical decisions. All relevant information should be weighted in, together with any appreciation of the costs of Type I and Type II errors (§21). H_T will rarely be exactly true and high variability, revealed by wide confidence intervals, may conceal a false H_T. *P*-values help other investigators to make their own assessments of the data.

§17. HYPOTHESIS TESTING FOR THE MEAN OF A NORMAL DISTRIBUTION: THE TWO-TAILED *t*-TEST

Suppose we have taken a random sample of n normally distributed observations, $x_1, x_2, \ldots, x_n$, and wish to test whether or not the sample supports the hypothesis that the value of μ_x is μ_T, where μ_T is some test value for μ_x which we ourselves have specified according to the objects of the investigation. Then, following the steps in §16, we proceed as follows:

1. **Hypothesis specification:** This can be stated as "the test hypothesis is that $\mu_x = \mu_T$, the alternative hypothesis is that μ_x is not equal to μ_T; we will operate with a maximum Type I error probability of α." The

investigator designates appropriate numerical values for μ_T and α and, abbreviating "is that" to ":", the specification is formally written

$$H_T: \mu_x = \mu_T, \quad H_A: \mu_x \neq \mu_T; \quad \alpha \quad (17.1)$$

The alternate hypothesis here is termed two-sided because it allows for μ_x-values either *greater or less than* μ_T. The one-sided H_A procedure is in §20. *The investigator decides*, in accordance with the experimental context, and *independently of the data outcome*, whether a one- or two-sided alternative hypothesis is appropriate.

2. **Test statistic:** For examining the significance of a normally distributed sample mean, the test statistic is defined to be

$$t_T = \frac{\bar{x} - \mu_T}{s_{\bar{x}}} \quad (17.2)$$

where $\bar{x} = \sum x/n$, $s_{\bar{x}} = \sqrt{s_x^2/n}$, and $s_x^2 = \sum' xx/(n-1)$. Since $\bar{x}$ will probably be close to μ_x, it will probably be close to μ_T if μ_T and μ_x *are* the same, in which case the deviation $|\bar{x} - \mu_T|$ and $|t_T|$ will probably be small. Conversely, if μ_T is far from μ_x, $|t_T|$ will probably be large. The denominator $s_{\bar{x}}$ standardizes the deviation to allow for the variability of $\bar{x}$ so that t_T is dimension-free.

3. **Assumption:** The procedure here relies on the normality assumption that $x_i \sim NI(\mu_x, \sigma^2)$, $i = 1, 2, \ldots, n$. It then follows, see (13.1), that, if H_T is true, $t_T = (\bar{x} - \mu_T)/s_{\bar{x}} = (\bar{x} - \mu_x)/s_{\bar{x}}$ is a member of the t-distribution with $(n-1)$ degrees of freedom.

4. **Critical region:** A *two-sided* H_A specification, indicated by the "$\neq$" in (17.1), signals a *two-tailed* critical region; H_T will be rejected and H_A accepted if t_T is either very large or very small. Accordingly, from the distribution tabulation of a random $t(n-1)$-variate we could read lower, t_L, and higher, t_H, critical values such that

$$P[t(n-1) < t_L] = P[t(n-1) > t_H] = \frac{\alpha}{2}$$

The total critical region comprises the abscissa points in the combined tails of the $t(n-1)$ distribution, from $-\infty$ to t_L and from t_H to $+\infty$ and, since t_T *is* a random $t(n-1)$-variate if H_T is true, the probability that t_T will fall in the critical region if H_T is true, thus leading us into a Type I error, is just $\alpha/2 + \alpha/2 = \alpha$ as specified. In practice, we exploit the symmetry and note that if t_T is more negative than t_L, $|t_T| = -t_T$ will exceed t_H which is just $t_H = t(n-1; \alpha/2)$. Our decision procedure is therefore to reject H_T and accept H_A if

$$|t_T| = |(\bar{x} - \mu_T)/\sqrt{s_x^2/n}| > t(n-1; \alpha/2) \tag{17.3}$$

and to stay with H_T if $|t_T| < t(n-1; \alpha/2)$.

5. **P-value:** If H_A is accepted, the probability that we have made a Type I error is $\leq \alpha$; we could write $P \leq \alpha$. More informatively, we could give at least an approximation to the *actual* probability of a Type I error as the probability that a random $t(n-1)$-variate will be more extreme than the test statistic value t_T realized and calculated. For a two-tailed critical region this is

$$P\text{-value} = 2P[t(n-1) > |t_T|] \tag{17.4}$$

The P-value will therefore decrease as $|t_T|$ increases.

Commonly, $|t_T|$ will fall between two of the numbers, in the $(n-1)$ degrees of freedom row, for which probabilities are given in the table heading. These can be specified or, if required, a generally sufficiently close approximation to the P-value may be obtained by linear interpolation.

6. **Confidence interval:** This may be calculated from (13.4). Alternatively, if, *but only if*, $\mu_T = 0$ and H_A is two-sided, the $100(1-\alpha)\%$ confidence interval for μ may be calculated as

$$\bar{x}\left(1 \mp \frac{t_{\alpha/2}}{t_T}\right) \tag{17.5}$$

Examples:

(i) From the random sample of $n = 4$ x-values 17, 16, 18, 21, test whether or not $\mu = 10$; take $\alpha = 0.05$. The analysis proceeds in steps as follows:

1. (17.1). H_T: $\mu = 10$, H_A: $\mu \neq 10$; $\alpha = 0.05$
2. (17.2). $\bar{x} = 74/4 = 18$, $\sum' xx = 14$, $s_x^2 = 14/3$, $s_{\bar{x}} = \sqrt{14/12}$.

$$t_T = \frac{18-10}{\sqrt{14/12}} = 7.4$$

3. Assuming $x_i \sim NI(\mu = 10, \sigma^2)$, $i = 1, \ldots, 4$, the test statistic is a member of the t-distribution with $n-1 = 3$ degrees of freedom.
4. In the 3 degrees of freedom row, beneath the 0.025 column of

Table A3, we find $t(3; 0.05/2) = 3.182$. The test statistic value $7.4 > 3.182$ and has fallen in the critical region. Hence, it being unlikely (though not impossible), that a t_3-variate more extreme than ∓ 3.182 would be obtained by chance if H_T were true, we make the statistical inference that the test hypothesis is false and prefer the alternative hypothesis. In so doing, we are accepting the risk, at a less than 1 in 20 chance, that we have been deceived by the data into making a Type I error.

5. (17.4). The test statistic value 7.4 is even greater than 5.841 found beneath 0.005 in the 3 degrees of freedom row. We accordingly simply write $P < 2(0.005) = 0.01$, noting that the actual probability of a Type I error is very small in this case.

6. (13.4). The 95% confidence interval for μ_x is

$$18 \mp (3.182)(\sqrt{14/12}) = 14.56, 21.44 \approx 14.5, 21.5$$

(ii) Suppose that, from a sample of $n = 25$ values, it is calculated that the sample mean is -1.75 and the sample variance, with $n - 1 = 24$ degrees of freedom, is $s^2 = 16.0$. Test the following:

1. H_T: $\mu = \mu_T = 0$, $\quad H_A$: $\mu \neq 0$; $\quad \alpha = 0.01$
2. The test statistic is $(-1.75 - 0)/\sqrt{16.0/25} = -2.188$.
3. Assuming the sample values are random and normally distributed, the test statistic is a random $t(24)$-variate if H_T is true.
4. From Table A3 we find $t(24; 0.01/2) = 2.797$ and, since $|t_T| = 2.188 < 2.797$, we accept the test hypothesis for $\alpha = 0.01$ in accordance with (17.3).
5. The test statistic is less decisive here since it can be checked that H_T would be rejected if $\alpha = 0.05$ had been specified because then $|t_T| = 2.19 > t_{0.025} = 2.064$. The P-value here directly measures the Type I error probability if we do decide to reject H_T. Linear interpolation between the t-values in the 24 degrees of freedom row which bracket $|t_T| = 2.188$ gives

$$P = 2\left[0.025 - \frac{(2.188 - 2.064)}{(2.492 - 2.064)}(0.025 - 0.01)\right] \approx 0.04$$

which does suggest the acceptance of the alternative hypothesis unless the reasons for taking $\alpha = 0.01$ are compelling.

6. The 99% confidence interval for μ is $-3.99, 0.49$ calculated as $-1.75 \mp 2.797(\sqrt{16/25})$ from (13.4). Alternatively, since (17.5) is now applicable, the interval can be calculated as $-1.75(1 \mp 2.797/2.188)$.

Notes: (a) If H_T is accepted we cannot make a Type I error. We might be wrong, however, in accepting H_T. This would be a Type II error (§21). (b) The fact that, in Example (ii), the 99% confidence interval for μ includes the test-value 0 for μ_T is consistent with the acceptance of H_T at $\alpha = 0.01$ (§19).

§18. HYPOTHESIS TESTING FOR THE VARIANCE OF A NORMAL DISTRIBUTION: THE TWO-TAILED χ^2-TEST

The §16 principles apply when hypotheses about variances of normal distributions are tested.

1. **Hypothesis specification:** Following §16, Step (1), we have

$$H_T: \sigma_x^2 = \sigma_T^2, \quad H_A: \sigma_x^2 \neq \sigma_T^2; \qquad \alpha \qquad (18.1)$$

where the investigator specifies the numerical values of σ_T^2 and α.

2. **Test statistic:** From the data, $x_1, \ldots, x_n$, we calculate s_x^2 and the test statistic is now

$$\chi_T^2 = \frac{(n-1)s_x^2}{\sigma_T^2} = \frac{\sum' xx}{\sigma_T^2} \qquad (18.2)$$

3. **Distribution:** From §14, if $x_i \sim NI(\mu, \sigma_x^2)$, $i = 1, \ldots, n$, then the quantity $\sum' xx/\sigma_x^2 \sim \chi^2(n-1)$. Hence, if H_T is true and $\sigma_x^2 = \sigma_T^2$, the test statistic is also a $\chi^2(n-1)$-variate; that is,

$$\frac{\sum' xx}{\sigma_T^2} \sim \chi^2(n-1)$$

so that, when H_T is true, tabulated $\chi^2(n-1)$ probabilities can be used as probabilities for $\sum' xx/\sigma_T^2$.

4. **Critical region:** From Table A4 for $f = n - 1$ degrees of freedom, we find a lower, $\chi^2(n-1; 1 - \alpha/2)$, and a higher, $\chi^2(n-1; \alpha/2)$, critical value such that

$$P[\chi^2 < \chi^2(n-1; 1-\alpha/2)] = \frac{\alpha}{2} = P[\chi^2 > \chi^2(n-1; \alpha/2)] \qquad (18.3)$$

The two-tail region, from 0 to $\chi_{1-(\alpha/2)}^2$ and from $\chi_{\alpha/2}^2$ to ∞, then comprises the critical region. We reject H_T in favor of H_A if the test statistic $\chi_T^2 = \sum' xx/\sigma_T^2$ falls in the region; we stay with H_T if $\chi_{1-(\alpha/2)}^2 < \chi_T^2 < \chi_{\alpha/2}^2$.

5. **P-Value:** The exceedance probability, assuming that H_T is true, of getting a random $\chi^2(n-1)$-variate more extreme than the test statistic χ_T^2 is taken in this, the two-tailed case, to be

$$2P[\chi^2(n-1) > \chi_T^2] \quad \text{or} \quad 2P[\chi^2(n-1) < \chi_T^2] \qquad (18.4)$$

for χ_T^2 in the right or left tail, respectively, of the $\chi^2(n-1)$-distribution.

6. **Confidence intervals:** Confidence intervals for σ_x^2 are calculated from (15.4).

Example: From a random sample of $n = 16$ normally distributed observations, it was calculated that $\sum' xx = \sum(x - \bar{x})^2 = 135$, so that $s_x^2 = 135/15 = 9$, with 15 degrees of freedom. At $\alpha = 0.05$, is this result statistically compatible with the hypothesis that $\sigma_x^2 = 20$?

1. $H_T: \sigma_x^2 = 20$, $H_A: \sigma_x^2 \neq 20$; $\alpha = 0.05$.
2. The test statistic (18.2) is $\chi_T^2 = \sum' xx/\sigma_T^2 = 135/20 = 6.75$.
3. If $x_i \sim NI(\mu, \sigma_x^2)$, $i = 1, \ldots, n = 16$, the test statistic is a $\chi^2(15)$-variate.
4. From the 15 degrees of freedom row of Table A4 we find $\chi^2(15; 0.975) = 6.262$ and $\chi^2(15; 0.025) = 27.488$. The test statistic value, $\chi_T^2 = 6.75$, lies between the two critical values and we accept H_T subject to the unknown but (we hope) small probability of having made a Type II error (§21).
5. Since $\mu[\chi^2(15)] = 15$ from §14, $\chi_T^2 = 6.75$ lies in the left tail of the $\chi^2(15)$-distribution and so the approximate P-value is obtained from (18.4) as $2P[\chi^2(15) < 6.75]$. By interpolation between the closest $\chi^2(15)$-values, $\chi_{0.975}^2 = 6.262$ and $\chi_{0.95}^2 = 7.261$, which bracket our χ_T^2, we first find

$$P[\chi^2(15) > 6.75] \approx 0.975 - \frac{(6.75 - 6.262)}{(7.261 - 6.262)}(0.975 - 0.95) = 0.96$$

and hence the exceedance probability is

$$2P[\chi^2(15) < 6.75] = 2(1 - 0.96) = 0.08$$

The probability of getting a test statistic value as or more extreme than 6.75, if H_T is true, is approximately 0.08.

§19. CONFIDENCE INTERVALS AND HYPOTHESIS TESTS RELATED

When the alternative hypothesis is two-sided and the exceedance probability is not required, confidence intervals can be used to give the outcomes of significance tests. The facts are that H_T is accepted for Type I error probability α, if the test value, μ_T for the mean or σ_T^2 for the variance, lies in the $100(1-\alpha)\%$ confidence interval for the parameter under test. The test hypothesis is rejected if the test value does not lie in the interval.

Population mean: By (17.3), H_T is accepted for Type I error probability α if, with $t_{\alpha/2} = t(n-1; \alpha/2)$,

$$\frac{|\bar{x} - \mu_T|}{\sqrt{s_n^2/n}} < t_{\alpha/2}, \quad \text{that is, if} \quad |\bar{x} - \mu_T| < t_{\alpha/2}\sqrt{\frac{s_x^2}{n}}$$

Accordingly, H_T is accepted if the equivalent inequalities hold that

$$\bar{x} - t_{\alpha/2}\sqrt{\frac{s_x^2}{n}} < \mu_T < \bar{x} + t_{\alpha/2}\sqrt{\frac{s_x^2}{n}}$$

in which case, the test value μ_T does lie in $100(1-\alpha)\%$ confidence interval for μ_x defined in (13.4).

Population variance: From §18(4) H_T is accepted if, with $n-1$ degrees of freedom

$$\chi_{1-(\alpha/2)}^2 < \frac{\sum' xx}{\sigma_T^2} < \chi_{\alpha/2}^2$$

that is, if

$$\frac{\sum' xx}{\chi_{\alpha/2}^2} < \sigma_T^2 < \frac{\sum' xx}{\chi_{1-(\alpha/2)}^2}$$

in which case σ_T^2 does lie in the confidence interval (15.4).

§20. TESTING ONE-SIDED ALTERNATIVE HYPOTHESES

In some practical contexts, one-sided rather than two-sided alternative hypotheses are appropriate. Thus, instead of $H_A: \mu_x \neq \mu_T$, we may require one, *not both*, of $H_A: \mu_x < \mu_T$ and $H_A: \mu_x > \mu_T$; instead of $H_A: \sigma_x^2 \neq \sigma_T^2$ we may require only one of $H_A: \sigma_x^2 < \sigma_T^2$ and $H_A: \sigma_x^2 > \sigma_T^2$.

relationship in §19 between the confidence interval and the significance test procedures does not apply now that H_A is one-sided. The analogous procedure for the other one-sided H_A case, H_A: $\mu_x > \mu_T$, is given as an example.

Example: With $\bar{x} = 12.5$ and $s_x^2 = 11/3$, three degrees of freedom, from $n = 4$ normally distributed observations, to test

1. H_T: $\mu_x \leq 10$, H_A: $\mu_x > 10$; $\alpha = 0.05$.
2. The test statistic is

$$\frac{\bar{x} - \mu_T}{s_{\bar{x}}} = \frac{12.5 - 10}{\sqrt{11/(3)(4)}} = 2.611.$$

3. Assuming that H_T is true and $x_i \sim NI(10, \sigma^2)$, $i = 1, \ldots, 4$, the test statistic is a random $t(3)$-variate.
4. Since H_A is "right-sided," the critical region comprises those $t(3)$-values exceeding $t_{0.05} = t(3; 0.05) = 2.353$ and since $t_T = 2.611 > 2.353$, we accept H_A.
5. The P-value, $P[t(3) > 2.611]$, is approximated by interpolation, using Table A3, as

$$P \approx 0.05 - \frac{(2.611 - 2.353)}{(3.182 - 2.353)} (0.05 - 0.025) \approx 0.04$$

6. Using (13.4) the 95% confidence interval for μ_x is

$$12.5 \mp 3.182\sqrt{11/12} = 9.4, \ 15.5$$

One-tailed tests for the population variance of a normal distribution:

1. According to the objectives of the tests, the hypothesis specification is one, but not both, of

$$H_T: \sigma_x^2 = \sigma_T^2, \quad H_A: \sigma_x^2 < \sigma_T^2; \qquad \alpha \qquad (20.5)$$

$$H_T: \sigma_x^2 = \sigma_T^2, \quad H_A: \sigma_x^2 > \sigma_T^2; \qquad \alpha \qquad (20.6)$$

2. In either case, the test statistic is

$$\chi_T^2 = \frac{(n-1)s^2}{\sigma_T^2} = \frac{\sum' xx}{\sigma_T^2} \qquad (20.7)$$

3. If $x_i \sim NI(\mu_x, \sigma_x^2 = \sigma_T^2)$, $i = 1, \ldots, n$, the test statistic is a $\chi^2(n-1)$-variate.

In such cases the principles in §16 still hold, although different, in one-tailed, critical regions are required.

One-tailed t-test for the population mean of a normal distribution: examine the possibility that $\mu_x < \mu_T$, for example, with Type I error probability α, the testing steps are as follows:

1. Hypothesis specification:

$$H_T: \mu_x \geq \mu_T, \quad H_A: \mu_x < \mu_T; \qquad \alpha \qquad (20.1)$$

where, as in the two-sided H_A case, the equality appears in the H_T part of the specification.

2. Test statistic:
From n observations, $\bar{x}$ and $s_{\bar{x}} = \sqrt{s_x^2/n}$ are calculated to give the test statistic t_T,

$$t_T = \frac{\bar{x} - \mu_T}{s_{\bar{x}}} = \frac{\bar{x} - \mu_T}{\sqrt{s_x^2/n}} \qquad (20.2)$$

3. Assumptions:
It is assumed that $x_i \sim NI(\mu_x, \sigma_x^2)$.

4. Critical region:
Small negative values of t_T could commonly arise by chance if H_T is true. If, however, t_T is extremely negative, it becomes prudent to prefer H_A. When H_T is true, t_T is a random $t(n-1)$-variate, for which

$$P[t(n-1) < -t(n-1; \alpha) = -t_\alpha] = \alpha$$

If we therefore reject H_T in favor of H_A whenever $t_T < -t_\alpha$, that is, falls in the critical region:

$$\text{the critical region is from } -\infty \text{ to } -t_\alpha = -t(n-1; \alpha) \qquad (20.3)$$

then the probability of being deceived into making a Type I error (the rejection of a true H_T) is just α. If $\bar{x} > \mu_T$, the test statistic (20.2) is necessarily positive so that H_T can be accepted "at sight."

5. P-Value:
The P-value is

$$P = P[t(n-1) < t_T] = P[t(n-1) > -t_T] \qquad (20.4)$$

6. Confidence interval:
The $100(1-\alpha)\%$ confidence interval for μ_x remains the same as that in (13.4), $\bar{x} \mp t_{\alpha/2}\sqrt{s_x^2/n}$ for the two-sided H_A case, using $t_{\alpha/2}$ and not the t_α used in (20.3). Correspondingly, the

4. For (20.5) the critical region is from 0 to $\chi^2_{1-\alpha} = \chi^2(n-1; 1-\alpha)$ because $P[\chi^2(n-1) < \chi^2_{1-\alpha}] = \alpha$. For (20.6) the critical region is from $\chi^2_\alpha = \chi^2(n-1; \alpha)$ to ∞. For example, H_A in (20.6) will be accepted if $\sum' xx/\sigma^2_T > \chi^2_\alpha$.

5. With χ^2_T as defined in (20.7) the P-values are $P[\chi^2(n-1) < \chi^2_T]$ for (20.5) and $P[\chi^2(n-1) > \chi^2_T]$ for (20.6).

Example: From $n = 16$ normally distributed observations it was calculated that $s^2_x = 9$. Was $\sigma^2_x > 5$?

1. Specification: $H_T: \sigma^2_x \leq 5$, $H_A: \sigma^2_x > 5$; 0.05.
2. Test statistic: $\chi^2_T = (n-1)s^2_x/\sigma^2_T = (15)(9)/5 = 27$.
3,4. Given normality, since $\chi^2_T = 27 > 24.996 = \chi^2(15; 0.05)$, the test statistic lies in the critical region. We "accept H_A."
5. By interpolation using the $\chi^2(15)$-tabulation, the P-value is 0.03.

Final Note: The specifications of H_A as two- or one-sided (and then which) and of numerical values for μ_T, σ^2_T, and α are made according to the contexts and objectives of particular investigations. Ideally, the choices should be made independently of, and not as a result of, data inspection and analysis, otherwise the probability assessments may need obscure revision. Thus, questions such as "Is treatment A definitely *better than* the control treatment?" and "Has this modification really *reduced* the variability?" invite one-sided H_A's. A two-sided H_A would be indicated for "Is treatment A *different from* the control treatment?".

§21. TYPE I AND TYPE II ERRORS AND POWER

Statistical tests do not prove or disprove hypotheses because sample to sample variability may give rise to a test statistic which deceptively favors H_A, even though H_T is true, and conversely. Inferences are made *on probability bases* in the hope of being much more often right than wrong. We recognize two ways of being wrong:

Type I error: The erroneous rejection of a true test hypothesis.

Type II error: The erroneous acceptance of a false test hypothesis.

The mnemonically arranged (A, False and Accept, alphabetically precede T, True and Reject, respectively) Table 21 presents the definitions

Table 21. Type I and Type II Errors Defined

Inference made	Actual situation	
	H_A is False	H_A is True
Accept H_A	Type I error	No error
Reject H_A	No error	Type II error

in terms of H_A. (It is a good exercise to make the corresponding table in terms of H_T.)

The probability of making a Type I error can be set at a chosen α-value; this α is referred to as the *size* of the test. *The probability (β) of making a Type II error cannot be specified* because the sample values used in calculating the test statistic depend on the *unknown* parameters of the *true H_A-distribution.* Thus, in testing a population mean, the test statistic (17.2) can be partitioned as

$$t_T = \frac{\bar{x} - \mu_T}{s_{\bar{x}}} = \frac{\bar{x} - \mu_x}{s_{\bar{x}}} + \frac{\mu_x - \mu_T}{s_{\bar{x}}} \qquad (21.1)$$

Given normality, the first term on the right-hand side is a straightforward $t(n-1)$-variate for which probabilities are tabulated. But, if H_A is true, $\mu_x - \mu_T \neq 0$ and, not knowing μ_x, ignorance of the second term in (21.1) prevents us from finding tail-area probabilities for the test statistic. Generally, however, we *can* say that the smaller we make α, the probability of a Type I error, the larger if H_A is true do we make β, the unknown probability of a Type II error, and conversely (see Fig. 21).

Two probability density curves are shown in Fig. 21:

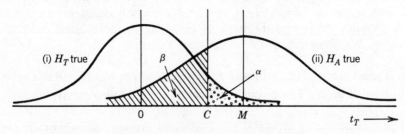

(i) H_T true β α (ii) H_A true

0   C   M t_T ⟶

Figure 21. α and β probabilities for H_T: $\mu_x = \mu_T$, H_A: $\mu_x > \mu_T$; α.

(i) That of the test statistic t_T when H_T is true.

(ii) That of the test statistic t_T when H_A: $\mu_x > \mu_T$ is true.

The distribution (i) is therefore the regular $t(n-1)$-distribution with population mean 0, as shown. The critical region for this right-sided H_A case comprises points on the t_T axis to the right of the point C, corresponding to the right-tail (dotted) area α. When H_A is true, the second term in (21.1) is positive and the mean of the t_T-distribution is accordingly shifted to the right, to the point M shown. The area under the curve (ii) to the right of the point C shows the probability that t_T will lie to the right of the point C, that is, will fall in the critical region and lead to the *correct decision*: H_A. The hatched area under curve (ii) to the left of the point C is the Type II error probability β, that t_T will not fall in the critical region accordingly leading to the *incorrect decision*: H_T. It can be seen that if the point C is moved to the right, to reduce α, the area β is increased, and conversely.

Power: The power, or sensitivity, of the test is measured by the probability $(1 - \beta)$ that a true H_A will be detected. Greater power, via a smaller β, can therefore be achieved by increasing α. Accordingly, in practice, α is advisedly chosen after a balanced assessment of the relative costs and consequences of Type I and Type II errors in the particular circumstances. Power can be increased, also at cost, by increasing the sample size n. This reduces $\sigma_{\bar{x}}$ and, via the second term in (21.1), increases the distance $0M$ in Fig. 21. An increase in sample size also reduces the standard deviations of the two distributions in Fig. 21, one effect of which is, for the same α, to contract the distance $0C$ thereby further enhancing power.

Comparing Two Groups

§22. SITUATIONS CONSIDERED

It is often required to compare the population means of two groups, treatments or varieties, using relatively small data samples. Different data generating procedures, such as different experimental designs and/or techniques, necessitate appropriately different data analyses. Some common situations to be considered are the following:

Situation	Analysis	Section
Randomized pairs data	Paired t-test	23
Completely randomized data		
(a) Group variances equal	Pooled variance t-test	24
(b) Group variances unequal	Approximate t-test	28
(c) Equality of group variances questionable	Variance ratio F-test before (a) or (b)	27

It will be assumed that the residual variability components are normally distributed variates. If this assumption appears dubious, a data transformation (§110) may be helpful and statistical texts may be consulted for alternative procedures such as distribution unspecified (nonparametric) analyses.

§23. RANDOMIZED PAIRS (RP): THE PAIRED t-TEST

Experimental plan and design: Observations are obtained on $2n$ experimental units grouped as n pairs, pair mates being genetically and/or temporally and/or spatially "close." To compare treatments A and B,

46

one unit *of each pair* is chosen at random to receive treatment A, the other unit receives treatment B. For high precision, pair mates should be as similar and similarly managed as possible except only for the A versus B difference being investigated. The precision of the intertreatment comparison is less and, ideally, not at all sensitive to appreciable differences *between* pairs.

Examples: (i) The comparison of two nutrition regimes using $n = 10$ pairs of identical twins. Fraternal twins give less precise paired comparisons. (ii) Sixteen potatoes divided into symmetrical halves to compare two cooking procedures ($n = 16$). (iii) Water quality measured upstream and downstream of $n = 15$ riparian cities.

Statistical model: For the observations or responses, y_{Ai} and y_{Bi}, recorded for treatments (groups, varieties) A and B, respectively, in the ith pair, we can write

$$y_{Ai} = \pi_i + \tau_A + \varepsilon_{Ai}, \qquad y_{Bi} = \pi_i + \tau_B + \varepsilon_{Bi}, \qquad i = 1, 2, \ldots, n \quad (23.1)$$

The responses, y_{Ai} and y_{Bi}, of both pair mates have a common component π_i representing the effect of the ith pair. We add τ_A for y_{Ai} and τ_B for y_{Bi} to represent the possibly different effects of the two treatments. Finally, regarding observations as population quantities afflicted by variability, we append individual residual variability or error components, ε_{Ai} and ε_{Bi}, to get more realistic constructions to model the actual observations y_{Ai} and y_{Bi}.

The statistical model guides us to the analysis; by subtraction, the model for the $(A - B)$ difference between the ith pair mates is

$$d_i = y_{Ai} - y_{Bi} = (\tau_A - \tau_B) + (\varepsilon_{Ai} - \varepsilon_{Bi}) \qquad (23.2)$$

The π_i terms have subtracted out; they do not affect the *within-pair* differences.

Now writing $(\tau_A - \tau_B) = \mu_d$ and $\varepsilon_{Ai} - \varepsilon_{Bi} = \delta_i$ and introducing the normality assumption for the δ_i variability element, we obtain the statistical model

$$d_i = \mu_d + \delta_i, \qquad \delta_i \sim NI(0, \sigma_d^2), \qquad i = 1, \ldots, n \qquad (23.3)$$

Apart from the notation change, d for x and δ for ε, (23.3) is exactly the model in (9.1) and hence the methods in §17 and §20 are immediately applicable to examine μ_d, the treatment difference of interest. In particular, $H_T: \mu_d = 0$ is equivalent to $H_T: \tau_A - \tau_B = 0$. The analyses, for two- and one-sided alternative hypotheses, respectively, are as follows.

48   COMPARING TWO GROUPS

Hypothesis specification, two-sided H_A:

$$H_T: \tau_A - \tau_B = \mu_T, \quad H_A: \tau_A - \tau_B \neq \mu_T; \qquad \alpha$$

or equivalently

$$H_T: \mu_d = \mu_T, \quad H_A: \mu_d \neq \mu_T; \qquad \alpha$$

where μ_T is a test value chosen according to the objectives of the investigation; for example, to examine whether or not $\tau_A = \tau_B$, we take $\mu_T = 0$.

Data: The data are the n differences,

$$d_1 = y_{A1} - y_{B1}, \quad d_2 = y_{A2} - y_{B2}, \quad \ldots, \quad d_n = y_{An} - y_{Bn}$$

Test statistic: By analogy with (17.2),

$$t_T = \frac{\bar{d} - \mu_T}{s_{\bar{d}}} \tag{23.4}$$

where

$$\bar{d} = \bar{y}_A - \bar{y}_B = \frac{\sum_{i=1}^n d_i}{n}, \quad s_{\bar{d}} = \sqrt{\frac{s_d^2}{n}}, \quad s_d^2 = \frac{\sum' dd}{n-1} \tag{23.5}$$

Distribution assumption: $\delta_i \sim NI(0, \sigma_d^2), \qquad i = 1, 2, \ldots, n.$

Critical region and inference: The two-tailed critical region is from $-\infty$ to $-t_{\alpha/2} = -t(n-1; \alpha/2)$ and from $t_{\alpha/2} = t(n-1; \alpha/2)$ to $+\infty$. H_T is rejected in favor of H_A if t_T falls in the critical region or equivalently, as in (17.3), if $|t_T| > t_{\alpha/2}$. Otherwise, we stay with H_T being aware of the unknown risk of a Type II error.

P-Value: Exactly as in (17.4) with t_T from (23.4)

$$P = 2P[t(n-1) > |t_T|] \tag{23.6}$$

Confidence interval: The $100(1-\alpha)\%$ confidence interval for the population mean difference, $\mu_d = \tau_A - \tau_B$, is calculated as

$$\bar{d} \mp t(n-1; \alpha/2) \sqrt{\frac{s_d^2}{n}} \tag{23.7}$$

where $\bar{d} = (\bar{y}_A - \bar{y}_B)$. The corresponding interval for $(\tau_B - \tau_A)$ is naturally

$$-\bar{d} \mp t(n-1; \alpha/2) s_{\bar{d}}$$

Hypothesis specification, one-sided H_A: The hypothesis specification is one of the two

$$\text{(i)} \quad H_T: \mu_d \leq 0, \quad H_A: \mu_d > 0; \quad \alpha$$

$$\text{(ii)} \quad H_T: \mu_d \geq 0, \quad H_A: \mu_d < 0; \quad \alpha$$

chosen according to the objectives of the investigation.

Data, test statistic, distribution assumption: These are the same as in the two-sided H_A situation above.

Critical region and inference: For specification (i) H_T is rejected in favor of H_A if the test statistic t_T falls in the right-tail critical region from $t(n-1; \alpha)$ to $+\infty$. For specification (ii) H_A is accepted if $t_T < -t(n-1; \alpha)$ or equivalently if $-t_T = |t_T| > t(n-1; \alpha)$.

P-Values: For specification (i), $P = P[t(n-1) > t_T]$; for specification (ii), $P = P[t(n-1) < t_T] = P[t(n-1) > |t_T|]$.

Confidence interval: The $100(1 - \alpha)\%$ confidence interval for $\tau_A - \tau_B$ is calculated as in the two-sided H_A case using (23.7).

Example: Suppose that diets A and B were compared using $n = 5$ pairs of cloned experimental animals. One animal, randomly chosen in each clone, received diet A, its clone mate received diet B. Suppose that the weight gains, in grams, were those shown in the A and B columns of Table 23 from which it is required to examine the population mean difference $(\tau_A - \tau_B)$. Assuming the statistical model $d_i = \mu_d + \delta_i, \quad \delta_i \sim$

Table 23. Data from a Randomized Pairs Experiment

Clone Pair	Weight Gains (g)		Difference (g), d
	A	B	
1	40	36	4
2	30	25	5
3	19	22	3
4	41	33	8
5	26	25	1
Total	156	141	15
Mean	31.2	28.2	3

$NI(0, \sigma_d^2)$, $i = 1, \ldots, 5$ and taking $\alpha = 0.05$, for example, we proceed as follows.

Hypothesis specification: H_T: $\mu_d = 0$, H_A: $\mu_d \neq 0$; 0.05.

Preliminary calculations: $\bar{d} = 3$ from Table 23, and

$$s_d^2 = \frac{\{(4^2 + 5^2 + \cdots + 1^2) - (15^2/5)\}}{4} = 17.5 \text{ (4 df)}$$

$$s_{\bar{d}} = \sqrt{\frac{s_d^2}{5}} = 1.871 \text{g}$$

Test statistic: $t_T = (3 - 0)/1.871 = 1.603$.

Critical region and inference: Since $t(n - 1; 0.025) = t(4; 0.025) = 2.776$ which exceeds 1.603, the test statistic has not fallen in the critical region. We accept H_T.

P-Value: The exceedance probability, $2P[t(4) > 1.603]$, is found to be 0.19 approximately [as in §17, Example (ii)]. Hence, if H_T were true, a test statistic as or more extreme than the one realized here could arise by chance in 1 out of 5 such experiments (in the long run). It seems prudent to accept H_T.

Confidence interval: The 95% confidence interval for $\tau_A - \tau_B$ is, from (23.7), or (17.5) since $\mu_T = 0$ here,

$$3 \mp 2.776(1.871) = -2.2, 8.2 \text{g}$$

Notes: We *have not proved* that $\tau_A = \tau_B$ but only that if $\tau_A \neq \tau_B$, the difference is too small to have been detected in this experiment. In fact, the confidence interval suggests that any value between -2.2 and 8.2g is a possibility for $\tau_A - \tau_B$. One such value in the interval, $\tau_A - \tau_B = 0$, is consistent with the outcome of the $\alpha = 0.05$ significance test as per §19. The wide confidence interval indicates an experiment of low precision and poor sensitivity for the detection of small differences.

§24. COMPLETELY RANDOMIZED (CR) DATA: TWO GROUPS

Experimental plan and design: Completely independent observations are obtained on a total of $(n_1 + n_2)$ experimental units; n_1 of the units, chosen at random, receive treatment (group, variety) A, while n_2 of the units receive B. All the units should be as similarly managed as possible except only for the A versus B difference being investigated. For a fixed total number of observations, precision is best when $n_1 = n_2$.

Examples: (i) A uniform field is subdivided into 16 equal plots. Eight plots, chosen at random from all the 16, receive fertilizer A, the remaining eight receive fertilizer B, so that $n_1 = n_2 = 8$ but, before harvest, hail damages two of the A and three of the B plots leaving $n_1 = 6$, $n_2 = 5$. (ii) To compare two cooking procedures on whole potatoes, 16 potatoes are divided into two random sets of eight. (iii) To examine air pollution in relation to city size, seven cities were selected at random from the small cities in a state. Their pollution concentrations were compared with those measured for seven randomly selected large cities ($n_1 = n_2 = 7$).

Statistical model: With y_{Ai} as the ith of the n_1 observations in group A and y_{Bj} as the jth of the n_2 observations in group B, the two sets of observations are modeled as

$$y_{Ai} \sim NI(\mu_A, \sigma^2), \qquad i = 1, \ldots, n_1$$
$$y_{Bj} \sim NI(\mu_B, \sigma^2), \qquad j = 1, \ldots, n_2$$

(24.1)

Equivalently, as in (9.1), the model is

$$y_{Ai} = \mu_A + \varepsilon_{Ai}, \qquad \varepsilon_{Ai} \sim NI(0, \sigma^2), \qquad i = 1, \ldots, n_1$$
$$y_{Bj} = \mu_B + \varepsilon_{Bj}, \qquad \varepsilon_{Bj} \sim NI(0, \sigma^2), \qquad j = 1, \ldots, n_2$$

(24.2)

Observations do not necessarily obey such simple models with additive, normally distributed, variability elements. Models are really assumptions which, we hope, are sufficiently close representations of the data generating procedures. The assumption that the population variance σ^2 is *the same for both groups* is particularly to be noted as a requisite for the analysis here. The normality assumption is not required until the critical region is defined for the probability assessment of the test statistic.

The model for the difference between the group means: Using (8.1) and (8.2), it follows from (24.1) that, for the sample means $\bar{y}_A$ and $\bar{y}_B$,

$$\bar{y}_A \sim N(\mu_A, \sigma^2/n_1) \quad \text{and} \quad \bar{y}_B \sim N(\mu_B, \sigma^2/n_2)$$

(24.3)

and since, as distinct from in the randomized pairs situation, the two sample means are now independent, we have from §7

$$V(\bar{y}_A - \bar{y}_B) = V(\bar{y}_A) + V(\bar{y}_B) = \frac{\sigma^2}{n_1} + \frac{\sigma^2}{n_2}$$

$$= \sigma^2 \left(\frac{1}{n_1} + \frac{1}{n_2} \right) = \frac{\sigma^2 (n_1 + n_2)}{n_1 n_2}$$

so that

$$(\bar{y}_A - \bar{y}_B) \sim N\{(\mu_A - \mu_B), \ \sigma^2 (n_1 + n_2)/n_1 n_2\} \qquad (24.4)$$

or equivalently,

$$(\bar{y}_A - \bar{y}_B) = (\mu_A - \mu_B) + (\bar{\varepsilon}_1 - \bar{\varepsilon}_2)$$

$$(\bar{\varepsilon}_1 - \bar{\varepsilon}_2) \sim N\{0, \ \sigma^2 (n_1 + n_2)/n_1 n_2\} \qquad (24.5)$$

This model is the basis for the analysis steps now presented for a two-sided H_A. If H_A is one-sided the steps are similar, except for the requirement of a correspondingly one-tailed critical region.

Hypothesis specification: The specification is written

$$H_T: \mu_A - \mu_B = \Delta_T, \quad H_A: \mu_A - \mu_B \neq \Delta_T; \qquad \alpha$$

where Δ_T is a test value for the $(\mu_A - \mu_B)$ difference, chosen according to the objectives of the investigation. For testing $H_T: \mu_A = \mu_B$ we take $\Delta_T = 0$.

Data: The data are the observations

$$y_{A1}, y_{A2}, \ldots, y_{An_1} \quad \text{and} \quad y_{B1}, y_{B2}, \ldots, y_{Bn_2}$$

Test statistic: The test statistic is

$$t_T = \frac{(\bar{y}_A - \bar{y}_B) - \Delta_T}{\sqrt{\hat{V}(\bar{y}_A - \bar{y}_B)}} = \frac{(\bar{y}_A - \bar{y}_B) - \Delta_T}{\sqrt{s^2 (n_1 + n_2)/n_1 n_2}} \qquad (24.6)$$

where s^2 is the estimate of the common variance σ^2.

Two independent, unbiased, estimates of σ^2 are available, one from each group. These are s_1^2 and s_2^2, where

$$s_1^2 = \frac{\sum (y_{Ai} - \bar{y}_A)^2}{n_1 - 1}, \quad \text{with } (n_1 - 1) \text{ df}$$

$$s_2^2 = \frac{\sum (y_{Bj} - \bar{y}_B)^2}{n_2 - 1}, \quad \text{with } (n_2 - 1) \text{ df}$$

These are combined or pooled to give their weighted average for the best, unbiased, estimate of σ^2 as

$$s^2 = \frac{(n_1-1)s_1^2 + (n_2-1)s_2^2}{n_1+n_2-2}$$

with $(n_1-1)+(n_2-1)=(n_1+n_2-2)$ df.

If the separate estimates s_1^2 and s_2^2 are not required, the preferred computation for s^2 is, using (2.8),

$$s^2 = \frac{\sum y_{Ai}^2 - (\sum y_{Ai})^2/n_1 + \sum y_{Bj}^2 - (\sum y_{Bj})^2/n_2}{n_1+n_2-2} \tag{24.7}$$

The numerical value of s^2 from (24.7) is substituted into (24.6) for the calculation of the test statistic.

Distribution assumptions: As noted under the statistical model above, it is being assumed that the two sets of data are random samples of independent observations from their respective normal distributions $N(\mu_A, \sigma^2)$ and $N(\mu_B, \sigma^2)$.

Critical region and inference: If H_T is true $(\mu_A - \mu_B)$ can be substituted for Δ_T in (24.6). With the normality assumption the test statistic then becomes a t-variate with (n_1+n_2-2) degrees of freedom. The critical region is from $-\infty$ to $-t(n_1+n_2-2; \alpha/2)$ and from $t(n_1+n_2-2; \alpha/2)$ to $+\infty$. Accordingly, H_T is rejected and H_A accepted if $|t_T| > t(n_1+n_2-2; \alpha/2) = t_{\alpha/2}$.

P-Value: $P = 2P[t(n_1+n_2-2) > |t_T|]$.

Confidence interval for $(\mu_A - \mu_B)$: The $100(1-\alpha)\%$ confidence interval for $(\mu_A - \mu_B)$ is, from (13.5),

$$(\bar{y}_A - \bar{y}_B) \mp t_{\alpha/2} \sqrt{\frac{s^2(n_1+n_2)}{n_1 n_2}} \tag{24.8}$$

If, but only if, $\Delta_T = 0$, the confidence interval may also be conveniently calculated as

$$(\bar{y}_A - \bar{y}_B)\left\{1 \mp \left(\frac{t_{\alpha/2}}{t_T}\right)\right\} \tag{24.9}$$

Confidence intervals for the common variance σ^2: These can also be obtained because

$$\frac{1}{\sigma^2}\sum(y_{Ai}-\bar{y}_A)^2 + \frac{1}{\sigma^2}\sum(y_{Bj}-\bar{y}_B)^2$$

is, from (14.3) with $f_1 = n_1 - 1$ and $f_2 = n_2 - 1$, a χ^2-variate with $(n_1 + n_2 - 2)$ degrees of freedom. From (15.4), a $100(1 - \alpha)\%$ confidence interval for σ^2 can therefore be obtained by dividing

$$\sum_1^{n_1} (y_{Ai} - \bar{y}_A)^2 + \sum_1^{n_2} (y_{Bj} - \bar{y}_B)^2 \qquad (24.10)$$

by $\chi^2_{\alpha/2} = \chi^2(n_1 + n_2 - 2; \alpha/2)$ and by $\chi^2_{1-(\alpha/2)}$, for the lower and upper limits, respectively.

Equal replication numbers: All the above procedures hold whether or not $n_1 = n_2$. If, however, $n_1 = n_2 = n$ say, the formulas become simpler since $V(\bar{y}_A - \bar{y}_B) = 2\sigma^2/n$. The test statistic (24.6) then becomes

$$t_T = \frac{(\bar{y}_A - \bar{y}_B - \Delta_T)}{\sqrt{2s^2/n}} \qquad (24.11)$$

where $s^2 = (s_1^2 + s_2^2)/2$ is obtained by dividing the numerator of (24.7) by the number of degrees of freedom, $2(n - 1)$. Critical values, P-values, and confidence intervals are correspondingly obtained using the t-distribution with $2(n - 1)$ degrees of freedom.

Example: The analysis of a typical two-group CR experiment with unequal numbers of replicate observations.

Hypothesis specification:

$$H_T: \mu_A = \mu_B, \quad H_A: \mu_A \neq \mu_B: \qquad \alpha = 0.05$$

Data:

Individual responses ($n_1 = 7$, $n_2 = 5$)

							Total	Mean
A	9.6,	9.9,	10.2,	9.9,	9.4,	10.5,	9.8	69.3	9.90
B	8.8,	8.9,	8.4,	8.6,	9.1,	—	—	43.8	8.76

Computations for the test statistic: Since,

$$s_1^2 = \frac{1}{6}\left\{(9.6)^2 + (9.9)^2 + \cdots - \frac{(69.3)^2}{7}\right\} = \frac{0.800}{6}, \quad 6 \text{ df}$$

$$s_2^2 = \frac{1}{4}\left\{(8.8)^2 + (8.9)^2 + \cdots - \frac{(43.8)^2}{5}\right\} = \frac{0.292}{4}, \quad 4 \text{ df}$$

the pooled variance estimate (24.7) is $s^2 = \{0.800 + 0.292\}/10 = 0.1092$,

10 df, and the test statistic (24.6) is

$$\frac{(9.90 - 8.76)}{\sqrt{0.1092(12/35)}} = 5.89$$

Inference: Assuming that the two sets of observations are random samples of independent observations from the normal distributions $N(\mu_A, \sigma^2)$ and $N(\mu_B, \sigma^2)$, the test statistic should, if H_T is true, be a random member of the t-distribution with 10 degrees of freedom. The two-tailed $\alpha = 0.05$ critical region consists of values more positive than $t(10; 0.025) = 2.228$ and values more negative than -2.228. The test statistic has fallen in this critical region and so supports the inference that $\mu_A \neq \mu_B$. If we accept this alternative hypothesis the associated probability of making a Type I error is $2P[t(10) > 5.89]$. This probability is not obtainable from Table A3 which, however, does show that $P < 2(0.005) = 0.01$.

Confidence interval for $(\mu_A - \mu_B)$*:* From (24.8) the 95% confidence interval for $(\mu_A - \mu_B)$ is

$$(9.90 - 8.76) \mp 2.228 \sqrt{0.1092 \left(\frac{12}{35}\right)} = 0.71,\ 1.57$$

or, alternatively from (24.9),

$$1.14 \left(1 \mp \frac{2.228}{5.89}\right) = 0.71,\ 1.57$$

The interval excludes 0; accordingly, 0 is not statistically compatible with the data as a value for $(\mu_A - \mu_B)$ at the 95% confidence level.

Confidence interval for σ^2*:* From the $\chi^2(10)$-table, $\chi^2(10; 0.025) = 20.48$, $\chi^2(10; 0.975) = 3.25$. The pooled numerator sum of squares (24.10) is $0.800 + 0.292 = 1.092$ so that the 95% confidence interval is

$$\frac{1.092}{20.48},\ \frac{1.092}{3.25} = 0.053,\ 0.336$$

Notes: (i) The above procedures for analyzing two-treatment, completely randomized data are special cases of the more general analysis of variance and F-distribution procedure (§70ff). (ii) In some cases s_1^2 and s_2^2 are so different that the equal variance assumption underlying the above analysis is dubious. The more complicated analysis then required begins in §27.

§25. WHICH DESIGN, THE RANDOMIZED PAIRS OR THE COMPLETELY RANDOMIZED, TO USE?

If $N = 2n$ available experimental units can be divided into pairs so that within-pair mates would give very close responses were they to receive the same treatment, then, despite perhaps appreciable differences between pairs, the RP design is preferable. One criterion for guiding the choice is the population mean of the semiwidth of the $100(1 - \alpha)\%$ confidence interval for the difference $(\mu_A - \mu_B)$ between the two means. On this basis, replacing the standard deviations by their population values in (23.7) and (24.8) and taking $n_1 = n_2 = n$ suggests that the RP design will be the better if

$$t(n - 1; \alpha/2) \frac{\sigma_d}{\sqrt{n}} < t(2n - 2; \alpha/2) \sqrt{\frac{2\sigma^2}{n}} \qquad (25.1)$$

Unless n is small, the difference between the two t-values will be minor so that (25.1) becomes approximately

$$\sigma_d < \sqrt{2\sigma^2}$$

This inequality will hold if strongly positively correlated pair mates can be obtained; see also §68 Example (ii). As another criterion for choice, the efficiency of an RP relative to a CR design can be estimated, with allowance for the degrees of freedom difference, as described in §108.

§26. F-DISTRIBUTIONS

Suppose we take two independent variates, $\chi^2(f_1)$ and $\chi^2(f_2)$, randomly drawn from χ^2-distributions with f_1 and f_2 degrees of freedom, respectively. If we divide the $\chi^2(f_1)$-variate by its number of degrees of freedom f_1, similarly divide $\chi^2(f_2)$ by f_2, and then form the ratio quantity, $[\chi^2(f_1)/f_1]/[\chi^2(f_2)/f_2]$, we have generated a random member of what is termed the $F(f_1, f_2)$-distribution, that one of the F-family of distributions which has f_1 numerator and f_2 denominator degrees of freedom. We write

$$\frac{\chi^2(f_1)/f_1}{\chi^2(f_2)/f_2} \sim F(f_1, f_2) \qquad (26.1)$$

Since, from §14, the population means of the $\chi^2(f_1)$ and $\chi^2(f_2)$ variates are just f_1 and f_2, respectively, the probabilities are high that the numerators and denominators in (26.1), and hence F-variate values, will be close to unity.

Example: If $y_i \sim NI(\mu, \sigma^2)$, $i = 1, \ldots, n$, then $\bar{y} \sim N(\mu, \sigma^2/n)$ and $(\bar{y} - \mu)^2/(\sigma^2/n)$ is a z^2-variate which, in turn, is a $\chi^2(1)$-variate, from §14. Furthermore, from §15, $\sum(y - \bar{y})^2/\sigma^2 = (n-1)s^2/\sigma^2$ is a $\chi^2(n-1)$-variate. Theory has shown that these two χ^2-variates are independent and hence the $F(1, n-1)$-variate obtained from (26.1) as

$$\frac{(\bar{y} - \mu)^2/(\sigma^2/n)}{(n-1)s^2/\sigma^2(n-1)} = \frac{(\bar{y} - \mu)^2}{s^2/n}$$

is, by (12.1), just the square of a $t(n-1)$-variate; a fact for later application.

Reading probability tables for F-variates: Values of $F(f_1, f_2)$-variates are related to right-tail area probabilities $p = P[F(f_1, f_2) > F(f_1, f_2; p)]$ in Table A5. The number of degrees of freedom f_1, in the numerator, is located in the row heading the table, the denominator degrees of freedom f_2 is located in the first column of the table to define the (f_1, f_2)-cell in which a selection of values $F(f_1, f_2; p)$ is given corresponding to the p-values in the second column of the table. This is illustrated in Fig. 26 for $f_1 = 7$, $f_2 = 8$, $p = 0.025$ for which $F(7, 8; 0.025) = 4.53$. A useful associated probability is also obtainable from the tabulation because, from the definition in (26.1), the reciprocal of an $F(f_1, f_2)$-variate is just an $F(f_2, f_1)$-variate. Hence, whenever $F(f_1, f_2)$ exceeds some number, the variate $F(f_2, f_1)$ will be less than the reciprocal of that number. Figure 26 therefore also gives the information that

$$p = 0.025 = P[F(7, 8) > 4.53] = P[F(8, 7) < 0.221]$$

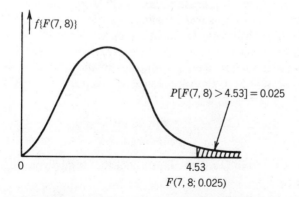

Figure 26. Illustration of F-variate tabulation.

As a general statement for this we have

$$P[F(f_1, f_2) > F(f_1, f_2; p)] = p = P[F(f_2, f_1) < 1/F(f_1, f_2; p)] \quad (26.2)$$

§27. TESTING FOR THE INEQUALITY (HETEROGENEITY) OF TWO VARIANCES

Suppose that $s_1^2, (n_1 - 1)$ df, and $s_2^2, (n_2 - 1)$ df, are the independent sample estimates of the variances σ_1^2 and σ_2^2, respectively, of two normally distributed populations and that it is required to test the two-sided H_A specification

$$H_T: \sigma_1^2 = \sigma_2^2, \quad H_A: \sigma_1^2 \neq \sigma_2^2; \quad \alpha$$

where α is the designated Type I error probability. Toward obtaining a test statistic we note from (15.1) that $(n_1 - 1)s_1^2/\sigma_1^2$ is a $\chi^2(n_1 - 1)$-variate, that $(n_2 - 1)s_2^2/\sigma_2^2$ is a $\chi^2(n_2 - 1)$-variate, and hence, by (26.1), that

$$\frac{(n_1 - 1)s_1^2/\sigma_1^2(n_1 - 1)}{(n_2 - 1)s_2^2/\sigma_2^2(n_2 - 1)} = \frac{s_1^2/\sigma_1^2}{s_2^2/\sigma_2^2} \quad (27.1)$$

is an $F\{(n_1 - 1), (n_2 - 1)\}$-variate.

On the usual presumption that H_T is true therefore, the population variances in (27.1) cancel and with the result that, for the variance ratio s_1^2/s_2^2,

$$s_1^2/s_2^2 \sim F\{(n_1 - 1), (n_2 - 1)\} \quad (27.2)$$

Accordingly, if H_T is true, we should expect s_1^2/s_2^2 to be close to unity and that either extremely small or extremely large values of s_1^2/s_2^2 suggest that H_T is false and H_A is true. Partitioning the α into $\alpha/2$ in the left-tail and $\alpha/2$ in the right-tail, the critical region comprises the abscissa points between 0 and $F\{(n_1 - 1), (n_2 - 1); 1 - (\alpha/2)\}$ and those from $F\{(n_1 - 1), (n_2 - 1); \alpha/2\}$ to $+\infty$. But, if s_1^2/s_2^2 lies in the left $\alpha/2$-"stretch," the reciprocal, s_2^2/s_1^2, lies in the right-tail $\alpha/2$-"stretch;" see (26.2). The test statistic is therefore taken as

$$F_T = \frac{\text{the larger } s^2}{\text{the smaller } s^2} \quad (27.3)$$

and H_T is rejected in favor of H_A, subject to a total Type I probability of $2(\alpha/2) = \alpha$, if

$$F_T > F\{f_{(L)}, f_{(S)}; \alpha/2\} \quad (27.4)$$

where $f_{(L)}$ and $f_{(S)}$ are the degrees of freedom for the larger and smaller variance estimates, respectively.

Example: To test for population variance inequality if the two variance estimates were $s_1^2 = 1.29$, 12 df, and $s_2^2 = 5.48$, 9 df.

Hypothesis specification: H_T: $\sigma_1^2 = \sigma_2^2$, H_A: $\sigma_1^2 \neq \sigma_2^2$;   0.05.

Test statistic: Since $s_2^2 > s_1^2$, $F_T = 5.48/1.29 = 4.248$.

Assumption: The variance estimates were calculated from normally distributed observations.

Critical region and inference: With $f_{(L)} = 9$ and $f_{(S)} = 12$ we read from Table A5 for 9 numerator and 12 denominator df and $p = 0.05/2 = 0.025$, that $F(9, 12; 0.025) = 3.44$. Since $4.248 > 3.44$ the test statistic has fallen in the critical region. We infer that $\sigma_1^2 \neq \sigma_2^2$, subject to a Type I error probability of 0.05.

§28. COMPLETELY RANDOMIZED DATA: TWO GROUPS, UNEQUAL VARIANCES

If from previous knowledge or from the test in §27 it is decided that the group population variances are not the same, the §24 procedure for testing the difference between two means needs modification. The statistical model (24.1) is changed to

$$y_{Ai} \sim NI(\mu_A, \sigma_1^2), \qquad i = 1, 2, \ldots, n_1$$
$$y_{Bj} \sim NI(\mu_B, \sigma_2^2) \qquad j = 1, 2, \ldots, n_2 \tag{28.1}$$

Hence, if we calculate s_1^2, from the A-observations, to estimate σ_1^2 and s_2^2, from the B-observations, to estimate σ_2^2, with $\sigma_1^2 \neq \sigma_2^2$, there is no common variance to be estimated by a pooled s^2. Some alternative procedures are (i) distribution unspecified or nonparametric analysis (Conover, 1980), (ii) data transformation (§110), (iii) assuming normality, the approximate test (Satterthwaite, 1946), (iv) the approximate test (Cochran and Cox, 1957), examined by Lauer and Han (1974), assuming normality, and (v) an exact procedure (Cox, 1985), assuming normality and that the variances are proportional to the population means or to their squares. Alternative (iii), which has been widely used, is now described.

Hypothesis specification: $H_T: \mu_A - \mu_B = \Delta_T, \quad H_A: \mu_A - \mu_B \neq \Delta_T; \quad \alpha.$

Data and assumptions: Statistics $\bar{y}_A, s_1^2; \bar{y}_B, s_2^2,$ calculated from samples according to the model (28.1) with $\sigma_1^2 \neq \sigma_2^2.$

Test statistic:

$$t_T^* = \frac{\bar{y}_A - \bar{y}_B - \Delta_T}{\sqrt{(s_1^2/n_1) + (s_2^2/n_2)}} \qquad (28.2)$$

Critical value and inference: H_T is rejected and H_A is accepted if $|t_T^*| > t_c = t(f; \alpha/2)$ where the "effective number of degrees of freedom," f, is calculated as

$$f \approx \frac{[(s_1^2/n_1) + (s_2^2/n_2)]^2}{\dfrac{(s_1^2/n_1)^2}{(n_1 - 1)} + \dfrac{(s_2^2/n_2)^2}{(n_2 - 1)}} \qquad (28.3)$$

To avoid interpolation, it will usually be satisfactory to round off the result from (28.3) to an integer value.

Confidence interval for $(\mu_A - \mu_B)$: An approximate $100(1-\alpha)\%$ confidence interval for $(\mu_A - \mu_B)$ can be calculated as

$$(\bar{y}_A - \bar{y}_B) \mp t(f; \alpha/2) \sqrt{\frac{s_1^2}{n_1} + \frac{s_2^2}{n_2}} \qquad (28.4)$$

Example: Suppose it is required to test $H_T: \mu_A = \mu_B, H_A: \mu_A > \mu_B; 0.01$ on the basis of the data statistics

$$\bar{y}_A = 7.82, \ s_1^2 = 5.48, \ n_1 = 9; \qquad \bar{y}_B = 4.93, \ s_2^2 = 1.29, \ n_2 = 12$$

From (28.2), with $\Delta_T = 0$ here, the test statistic is

$$t_T^* = \frac{7.82 - 4.93}{\sqrt{\dfrac{5.48}{9} + \dfrac{1.29}{12}}} = 3.42$$

For the effective number of degrees of freedom, (28.3) gives

$$\frac{[(5.48/9) + (1.29/12)]^2}{\dfrac{(5.48/9)^2}{8} + \dfrac{(1.29/12)^2}{11}} = 10.83$$

Noting that H_A is one-sided, we read $t(10; 0.01) = 2.764, \ t(11; 0.01) =$

2.718, so that $t_T^* = 3.42$ is in the critical region whether we round down or up. We can infer that $\mu_A > \mu_B$ and $P < 0.005$.

From (28.4) using $f = 10$ and $t(10; 0.005) = 3.169$, the approximate 99% confidence interval for $(\mu_A - \mu_B)$ is

$$(7.82 - 4.93) \mp 3.169 \sqrt{\frac{5.48}{9} + \frac{1.29}{12}} = 0.2,\ 5.6$$

Notes: (i) The standard procedure is inaccurate in cases when $\sigma_1^2 \neq \sigma_2^2$ because, although when H_T is true the test statistic (28.2) does have the general t-variable form, theory has established that t_T^* is not a "proper" t-variate when H_T is true. (ii) Strictly, if the procedure here follows the variance inequality "preliminary test," the inferences about $(\mu_A - \mu_B)$ depend on the assumption that the $\sigma_1^2 \neq \sigma_2^2$ inference is correct—but it may not be. This "inference with conditional specification" situation is discussed in Bancroft and Han (1981) with some suggestions on apposite Type I error probabilities in the overall procedure.

Some Discrete, Categorized, Data Procedures

§29. INVESTIGATION CONTEXTS

Inferences from discrete data are often required when, instead of being measurements of continuous variates, the basic data are counts of individuals in categories. Examples arise in studies of pre-election polls, of divorce incidence cross-classified by socioeconomic level and religious affiliation, and of (Mendel) sweet-peas classified by color. To set the stage for some available analysis procedures, two probability axioms are introduced.

§30. PROBABILITY, INDEPENDENT EVENTS, AND TWO PROBABILITY AXIOMS

Probability: In discrete data situations, the variate must take one of a set of distinct values, an occurrence conveniently termed an event, E. Thus, a rolled die might give the particular event $E_{(6)}$: "the uppermost number is 6." If a specified event occurs r times in n independent opportunities, the probability of the event is estimated as the proportion r/n. As an intuitive abstraction from practical experiences, we take $P(E)$, the *probability* for the occurrence of a specified event E, as the *long-run limit* of the estimating proportion, as the number of opportunities increases indefinitely.

Independent events: Two events, E_1 and E_2, are statistically independent if the probability that one occurs is in no way dependent on the probability that the other occurs.

62

Example: Take E_1 as the answer "Yes" when a person is asked "Do you like cats?" and E_2 as the answer "Yes" when another person is asked "Do you like dogs?". If unrelated respondents are randomly chosen from different households, E_1 and E_2 should be independent. Will the events be independent if the two respondents are living together?

Probability axiom 1. Independent events: If the events $E_1, E_2, \ldots, E_m$ are all *independent, the probability that they all occur is the product* of the individual probabilities $P(E_1), P(E_2), \ldots, P(E_m)$. Thus,

$$P(E_1 \text{ and } E_2 \text{ and } \cdots \text{ and } E_m) = P(E_1)P(E_2) \cdots P(E_m) \quad (30.1)$$

Example: A die is rolled independently two times or two dice, randomly arranged, are rolled. The probability of getting two sixes is $(1/6)(1/6) = 1/36$.

Mutually exclusive events: If, of several events, $E_1, E_2, \ldots, E_m$, only one can occur on any given occasion, the events are mutually exclusive or disjoint.

Example: $E_1 = $ a six on a rolled die, $E_2 = $ a five on a rolled die, $E_3 = $ an even number on a rolled die. E_1 and E_2 are disjoint events, E_2 and E_3 are disjoint events, and E_1 and E_3 are not disjoint events.

Probability axiom 2. Disjoint events: If the events $E_1, E_2, \ldots, E_m$ are *disjoint, the probability that one of them occurs is the sum* of the probabilities $P(E_1), \ldots, P(E_m)$. Thus,

$$P(E_1 \text{ or } E_2 \text{ or } \cdots \text{ or } E_m) = P(E_1) + P(E_2) + \cdots + P(E_m) \quad (30.2)$$

Example: $P(|z| > 1.96) = P(z < -1.96 \text{ or } z > +1.96)$. The two events $z < -1.96$ and $z > +1.96$ are mutually exclusive. Hence,

$$P(z < -1.96 \text{ or } z > 1.96) = P(z < -1.96) + P(z > 1.96)$$

$$= 2P(z > 1.96) = 0.05$$

If, in addition to being mutually exclusive, the events $E_1, E_2, \ldots, E_m$ are *exhaustive* in that it is certain that *one of them must occur*, then

$$P(E_1 \text{ or } E_2 \cdots \text{ or } E_m) = \sum_1^m P(E_i) = 1$$

Thus, in rolling a die,

$$P(1 \text{ or } 2 \text{ or } \cdots \text{ or } 6) = 6(1/6) = 1$$

§31. BINOMIAL DISTRIBUTIONS: DEFINITION

A binomial distribution may be recognized by the following three attributes:

1. An experiment or investigation consists of a known number, n, of independent trials.
2. The outcome of each trial must be one of two, only, mutually exclusive events E_1 and E_2. Generically, we term these as $E_1 =$ success and $E_2 =$ failure. For example, (i) if "cured" is success and not-cured is failure; (ii) if female is success and male is failure(!); (iii) if a rolled die showing a 6 is success and a "not-6," that is, any one of $1, \ldots, 5$, is failure.
3. The probability of success, π say, is constant for all the independent trials—so also then is the probability of failure, $(1 - \pi) = Q$ say—there being no Greek letter Q.

In repetitions of the n-trial experiment, different integer values will usually be obtained for the number, x say, of successes. Accordingly, x is a discrete variate. We say that the variate x "is binomially distributed" or "has a binomial distribution." The distribution is determined by the numerical values of its two parameters n and π. We write

$$x \sim b(n, \pi) \tag{31.1}$$

The probability of failure is $(1 - \pi)$ so that, if $y = n - x$ is the number of failures, $x \sim b(n, \pi)$ implies that $y \sim b(n, 1 - \pi)$.

The "observation" from any one experiment is the value, r say, of the number-of-successes, discrete, variate x and, to find the *distribution* of the variate, we need the probability $P(x = r)$ for each possible value of x, that is, for $r = 0, 1, \ldots, n$. For $r = 0$ we have

$$P(x = 0) = P(\text{failure and failure} \cdots \text{and failure})$$

and if $P(\text{failure}) = 1 - P(\text{success}) = (1 - \pi)$, this gives, by (30.1),

$$P(x = 0) = (1 - \pi)^n \quad \text{and similarly} \quad P(x = n) = \pi^n$$

With $Q = 1 - \pi$, the probabilities $P(x = r)$ are the successive terms in the binomial expansion

$$1 = (\pi + Q)^n = Q^n + n\pi Q^{n-1} + \cdots + \binom{n}{r} \pi^r Q^{n-r} + \cdots + \pi^n \tag{31.2}$$

wherein $\binom{n}{r}$, read as "n choose r," is the number of different ways of choosing r things from a group of n things, expressed as

$$\binom{n}{r} = \frac{n!}{r!(n-r)!} = \frac{n(n-1)\cdots(n-r+1)}{r(r-1)\cdots 3\cdot 2\cdot 1} \tag{31.3}$$

after canceling, because $i!$, read as "i factorial," is the symbolic representation of the product

$$i! = i(i-1)(i-2)\cdots 3\cdot 2\cdot 1$$

which appears for $i = n$, $i = r$, and $i = (n-r)$ in (31.3). Thus, for $i = 5$, $5! = 5\cdot 4\cdot 3\cdot 2\cdot 1 = 120$. By convention, $0!$ is taken to be unity. From (31.2) then, the probability of getting exactly $x = r$ successes in an experiment consisting of n independent binomial trials is

$$P(x = r) = \binom{n}{r}\pi^r(1-\pi)^{n-r}, \qquad r = 0, 1, \ldots, n \tag{31.4}$$

Example: What are the probabilities for the alternative numbers of 6's when a die is rolled, independently, four times? Designating a 6 as a success and anything else, a not-6, as a failure, we seek the distribution of the $x =$ the number of successes, when $x \sim b(4, 1/6)$. From (31.4) with $n = 4$, $\pi = 1/6$,

$$P(x = 0) = \frac{4!}{0!\,4!}\left(\frac{1}{6}\right)^0\left(\frac{5}{6}\right)^4 = \left(\frac{5}{6}\right)^4 = 0.482$$

$$P(x = 1) = \frac{4!}{1!\,3!}\left(\frac{1}{6}\right)^1\left(\frac{5}{6}\right)^3 = 4\left(\frac{1}{6}\right)\left(\frac{5}{6}\right)^3 = 0.386$$

$$P(x = 2) = \frac{4!}{2!\,2!}\left(\frac{1}{6}\right)^2\left(\frac{5}{6}\right)^2 = 6\left(\frac{1}{6}\right)^2\left(\frac{5}{6}\right)^2 = 0.116$$

Continuing similarly for $r = 3$ and 4 enables the complete distribution, showing the alternative x-values together with their associated probabilities, to be presented as in Table 31.

For a continuous variate, the total area under the probability density function curve is unity; the discrete variate analogy is that the sum of the probabilities for all the various values of the variate is also unity from

Table 31. The Binomial, $n = 4$, $\pi = 1/6$, Distribution

r	0	1	2	3	4
$P(x = r)$	0.482	0.386	0.116	0.015	0.001

(31.2) and as in Table 31 where

$$\sum_{r=0}^{r=4} P(x = r) = 1$$

§32. BINOMIAL DISTRIBUTION: POPULATION MEAN AND VARIANCE

If $x \sim b(n, \pi)$ and we abbreviate $P(x = r)$ to $P(r)$, the facts are:

The population mean of x is $\mu_x = n\pi$

(32.1)

The population variance of x is $V(x) = n\pi(1 - \pi) = n\pi Q$

The results (32.1) can be proved algebraically from the basic definitions

$$\mu_x = \sum_{r=0}^{n} rP(r) \quad \text{and} \quad V(x) = \sum_{r=0}^{n} (r - \mu_x)^2 P(r)$$

the latter, for example, being analogous to the definition of population variance for a continuous variate (5.1); the former may also be compared to the calculation (2.2) for a sample mean using relative frequencies.

§33. PROPORTIONS AND PERCENTAGES

When x successes are observed in n independent binomial trials, practical interest often centers on the proportion of successes, $p = x/n$, of which the various alternative values are $0/n, 1/n, \ldots (n-1)/n, n/n = 1$. The probability that x/n takes the value r/n is just $P(x = r)$ so that, from (31.4),

$$P(p = r/n) = P(x/n = r/n) = \binom{n}{r} \pi^r (1 - \pi)^{n-r} \tag{33.1}$$

For the distribution of the $p = x/n$ proportion-variate, the facts are:

The population mean of x/n is $\mu_p = \mu[x/n] = \pi$

The population variance of x/n is $\sigma_p = V(x/n) = \pi(1 - \pi)/n$ (33.2)

The population standard deviation of x/n is $\sqrt{\pi(1 - \pi)/n}$

The sample estimate of π is $\hat{\pi} = p = x/n$

The sample estimate of $V(x/n)$ commonly used is $p(1 - p)/n$; the unbiased estimate is $p(1 - p)/(n - 1)$ (33.3)

Note: (i) Since $x/n = (1/n)x$, the results (33.2) follow immediately from (32.1) using Rule 1 (6.2) with $d = 0$ and $c = 1/n$. (ii) Rule 1 can also be applied to obtain results for percentages. Thus, the population mean of the percentage $100x/n$ is 100π and the population variance is $100^2 \pi(1 - \pi)/n$.

§34. NORMAL DISTRIBUTION APPROXIMATIONS FOR BINOMIAL DISTRIBUTIONS

Many analysis procedures for binomially distributed data use the theoretically established result that, as n increases, binomial distributions become progressively more like normal distributions. In particular, when n is large enough x, the number of successes in n trials, has approximately that normal distribution with population mean $n\pi$ and population variance $n\pi(1 - \pi)$ as in (32.1) and, equivalently, the proportion $p = x/n$ has approximately that normal distribution with population mean π and population variance $\pi(1 - \pi)/n$ as in (33.2). Thus, if $x \sim b(n, \pi)$ and n is large enough, we have approximately

$$x \sim N\{n\pi, n\pi(1 - \pi)\} \quad \text{and} \quad p \sim N\{\pi, \pi(1 - \pi)/n\} \qquad (34.1)$$

Binomial distributions are already symmetrical if $\pi = 0.5$, and if π is close to 0.5, good agreement between the binomial and its approximate normal distribution is obtained for smaller n-values than those required when π is close to 0 or 1. Wetherill (1982) suggested that for the approximation to be reasonable

$$n \text{ should exceed the larger of } \frac{9\pi}{(1 - \pi)} \text{ and } \frac{9(1 - \pi)}{\pi} \qquad (34.2)$$

and be considerably larger if tail-area probabilities <0.01 are of interest.

§35. HYPOTHESIS TESTS AND INTERVAL ESTIMATIONS FOR A SINGLE PROPORTION

The binomial variate analogues of the continuous variate situations in §13, §17, and §20 arise when we seek information about a population *proportion* π, instead of a population *mean* μ, using the observed proportion $\hat{\pi} = p = x/n$, instead of the sample mean $\hat{\mu} = \bar{x}$. We look separately at two situations according to whether n is not, or is, large enough for a satisfactory normal distribution approximation.

1. **Examining proportions when n is small:** Suppose that the datum consists of the observation of x successes in n independent trials, that $x \sim b(n, \pi)$, where π is the unknown population proportion in which we are interested, but that n is not large enough for the binomial distribution of x to be approximated by a normal distribution. Suppose it is desired to examine the following:

Hypothesis specification:

$$H_T: \pi \le \pi_T, \quad H_A: \pi > \pi_T \tag{35.1}$$

where the test value π_T, is specified by the investigator. We do not calculate here a simple test statistic and check to see if it lies in a critical region defined according to a prescribed Type I error probability α, equal to 0.05, for example. The reason is that, because x is now discrete, the tail-area probabilities do not change continuously but in "jumps" as in Table 31. Instead, we presume, as usual, that $\pi = \pi_T$ and proceed to calculate the right-tail exceedance probability P of obtaining x or more successes or, equivalently, of obtaining a proportion of successes equal to or greater than that observed, $p = x/n$. If P is unreasonably small on the presumption that H_T is true, we reject H_T in favor of H_A and conversely. Hence, we find

$$P = P[x \text{ or } (x+1) \text{ or } \cdots \text{ or } n, \text{ successes}] = \sum_{r=x}^{n} \binom{n}{r} \pi_T^r (1 - \pi_T)^{n-r} \tag{35.2}$$

using (30.2) and (31.4). Alternatively, we can use the strange but theoretically established result that, in terms of an F-variate,

$$P = P[F\{2(n+1-x), 2x\} > x(1 - \pi_T)/(n+1-x)\pi_T] \tag{35.3}$$

Example: Suppose we have bet a professional die roller that (s)he cannot roll four or more 6's out of eight rolls and become disconcerted when (s)he does, in fact, roll four 6's. Discounting psychokinesis, we may wonder if some jiggery-pokery is involved. Specifically, we wish to examine

$$H_T: \pi = \frac{1}{6}, \quad H_A: \pi > \frac{1}{6}$$

From (35.2) we can calculate

$$P = \sum_{r=4}^{8} \binom{8}{r} \left(\frac{1}{6}\right)^r \left(\frac{5}{6}\right)^{8-r} = 0.0306$$

or, and especially when (35.2) entails excessive effort, from (35.3) we have, with $n = 8$ and $x = 4$,

$$P = P[F(10, 8) > 4(5/6)/5(1/6)] = P[F(10, 8) > 4]$$

Computer programs are available to obtain such probabilities; alternatively, Table A5 shows that $0.025 < P < 0.05$ and linear interpolation gives $P \approx 0.033$. By either procedure H_A is indicated. If, instead of (35.1), we wish to test $H_T: \pi \geq \pi_T$, $H_A: \pi < \pi_T$, on the basis of x successes in n trials, we can test the equivalent specification in terms of the number of failures; that is, we test

$$H_T: (1 - \pi) \leq (1 - \pi_T), \quad H_A: (1 - \pi) > (1 - \pi_T)$$

on the basis of $(n - x)$ failures observed in n trials. If a two-sided H_A is specified, it is usual to take the exceedance probability as twice that calculated above.

Confidence limits for the population proportion: Tables are available to obtain confidence intervals for a selection of x and n values. Alternatively, we can again use F-distribution probabilities in which case the $100(1 - \alpha)\%$ confidence interval for π is from

$$\frac{x}{x + (n + 1 - x)F_1} \quad \text{to} \quad \frac{(x + 1)F_2}{n - x + (x + 1)F_2} \tag{35.4}$$

where

$$F_1 = F\{2(n + 1 - x), 2x; \alpha/2\} \quad \text{and} \quad F_2 = F\{2(x + 1), 2(n - x); \alpha/2\} \tag{35.5}$$

Example: Find the 95% confidence interval for π if $x = 15$ successes are observed out of $n = 20$ trials. With $F_1 = F\{12, 30; 0.025\} = 2.41$ and $F_2 = F\{32, 10; 0.025\} \approx 3.30$, (35.4) gives the interval from

$$\frac{15}{15 + 6(2.41)} = 0.51 \quad \text{to} \quad \frac{16(3.30)}{5 + 16(3.30)} = 0.91$$

Note: Angus and Shafer (1984) have shown that the above intervals, based on Clopper and Pearson (1934), correspond to slightly more than the nominal $100(1 - \alpha)\%$ prescribed confidence level; that is, the intervals are slightly larger than they should be. The excesses, however, do not appear to be alarmingly large for $n > 10$.

2. **Examining proportions when n is large:** The F-distribution procedures given above are applicable, in fact, when n is large. In this case,

however, the simple approximate procedure based on the approximate normal distributions for x or $p = x/n$ is commonly used. In what follows, the general procedure will be concurrently illustrated in terms of an example.

Example: To see if people were in favor of a land-use bill, a random sample of 240 people was taken; 140 favored the bill. Does this support the hypothesis that more than half of the population sampled favored the bill?

Hypothesis specification; Taking a one-sided H_A case, the specification is

$$H_T: \pi = \pi_T, \quad H_A: \pi > \pi_T; \quad \alpha$$

so that, in the example, taking $\pi_T = \frac{1}{2}$ and $\alpha = 0.01$, we have

$$H_T: \pi = \frac{1}{2}, \quad H_A: \pi > \frac{1}{2}; \quad 0.01$$

Data: x successes are observed in n trials to give $\hat{\pi} = p = x/n$. In the example we have $\hat{\pi} = p = 140/240 = 0.583$.

Assumptions: We assume that x, and correspondingly p, have binomial distributions; in particular, that $p \sim b(n, \pi)$ and that n is sufficiently large to support procedures based on the approximation (34.1) that $p \sim N\{\pi, \pi(1-\pi)/n\}$. Accordingly, in the example, we assume that $p \sim N(0.5, 0.25/240)$, having noted that the criterion (34.2) is well satisfied for $\pi_T = 1 - \pi_T = 1/2$, because $240 > 9$.

Test statistic: The test statistic is

$$z_T = \frac{x - n\pi_T}{\sqrt{n\pi_T(1 - \pi_T)}} = \frac{p - \pi_T}{\sqrt{\pi_T(1 - \pi_T)/n}} \qquad (35.6)$$

It is labeled z_T because, if H_T is true, each denominator is the *population standard deviation* of the corresponding numerator deviation, in accordance with the definition of a standard normal z-variate. For the example we find

$$z_T = \frac{0.583 - 0.5}{\sqrt{(0.5)(1 - 0.5)/240}} = 2.57$$

Critical region and inference: The critical region is that for a probability α in the right-tail of the z-distribution, so that H_T is rejected if $z_T > z_\alpha$.

For $\alpha = 0.01$, $z_{0.01} = 2.326$ and, since $2.57 > 2.33$, H_T is rejected; the indication is that the population proportion in favor of the bill was greater than $1/2$.

Exceedance probability: The P-value here is $P(z > z_T)$ and from the z-table we read $P = P(z > 2.57) = 0.005$ as the actual probability that the acceptance of H_A entails a Type I error.

Confidence interval: The $100(1 - \alpha)\%$ confidence interval for π is

$$p \mp z_{a/2} \sqrt{\frac{p(1-p)}{n}} \qquad (35.7)$$

with the apparent inconsistency that $z_{\alpha/2}$, not $t_{\alpha/2}$, is used even though $p(1-p)/n$ is the estimate of $V(p)$ and not the actual population variance. The form (35.7) is used even if H_T is accepted for it is not then *proved* that $\pi = \pi_T$. In the example, the 95% confidence interval is, with $z_{\alpha/2} = z_{0.025} = 1.96$,

$$0.583 \mp 1.96 \sqrt{\frac{(0.583)(0.417)}{240}} = 0.52, 0.65$$

and, correspondingly, the 95% confidence interval for the *population percentage*, 100π, is from 52 to 65%.

Note: Continuity correction: The Yates (1934) "continuity correction" is sometimes used in the expressions for calculating the test statistic and confidence interval. The correction always makes the test statistic less extreme. The correction has little effect when n is large and is not used here in agreement with Fienberg (1977) where details and references on the correction are given.

Confidence intervals for a proportion when no successes are observed in n trials: In this special but sometimes highly important case the, termed "one-sided," $100(1 - \alpha)\%$ confidence interval for π can be calculated (e.g., see Louis, 1981), as from

$$0 \quad \text{to} \quad 1 - (\alpha)^{1/n}$$

and, correspondingly, if n successes are observed in n trials, the $100(1 - \alpha)\%$ confidence interval is from $(\alpha)^{1/n}$ to 1. Thus, if 0 successes are observed in $n = 8$ trials, the 95% confidence interval for π is from 0 to $1 - (0.05)^{1/8} = 0.31$ while, if 8 successes out of 8 trials are observed, the 95% confidence interval for the population success percentage is from $100(0.05)^{1/8} = 69\%$ to 100%.

§36. COMPARING TWO INDEPENDENT SAMPLE PROPORTIONS

As the discrete variate analogue of the §24 continuous variate situation, we assess here the sample evidence respecting the equality or otherwise of two population proportions, π_1 for group 1 and π_2 for group 2.

Hypothesis specification: Taking a two-sided H_A case, we specify

$$H_T: \pi_1 = \pi_2, \quad H_A: \pi_1 \neq \pi_2; \quad \alpha$$

Data:

Group	Successes	Opportunities	Proportions
1	x_1	n_1	$x_1/n_1 = p_1$
2	x_2	n_2	$x_2/n_2 = p_2$

Assumptions: For very small sample sizes, it is advisable to use Fisher's exact test based on exceedance probabilities. The test is easily carried out using specially prepared tables; for example, those in McGuire et al. (1967). We assume here that both samples are large enough to sustain normal distribution approximations. On this, one empirical recommendation is that $(n_1 + n_2)$ should generally exceed 20 and should exceed 40 if either x_1 or x_2 is less than 5. Alternatively, the less stringent criterion in §40, Assumption, may be adopted. Specifically then, we assume here that

$$p_i \sim N(\pi_i, \pi_i Q_i / n_i), \quad Q_i = (1 - \pi_i), \quad i = 1, 2 \qquad (36.1)$$

and that the two proportions are independent.

Test statistic: The test statistic is $(p_1 - p_2)/\sqrt{\hat{V}(p_1 - p_2)}$ where, under the assumptions (36.1),

$$\mu(p_1 - p_2) = \pi_1 - \pi_2$$

$$V(p_1 - p_2) = V(p_1) + V(p_2) = \frac{\pi_1 Q_1}{n_1} + \frac{\pi_2 Q_2}{n_2}$$

Now presuming that H_T is true, with $\pi_1 = \pi_2 = \pi$ say, gives $\mu(p_1 - p_2) = 0$ and

$$V(p_1 - p_2) = \frac{\pi Q}{n_1} + \frac{\pi Q}{n_2} = \frac{\pi Q(n_1 + n_2)}{n_1 n_2}$$

Furthermore, the natural estimate of a common population proportion π is $\hat{\pi} = p = (x_1 + x_2)/(n_1 + n_2)$, obtained from the combined samples. Replacing π and Q in the variance formula by this p and $q = 1 - p$ and taking the square root, we obtain the test statistic

$$z_T = \frac{p_1 - p_2}{\sqrt{pq(n_1 + n_2)/n_1 n_2}} \qquad (36.2)$$

of which a convenient, algebraically equivalent, expression is

$$z_T = (n_2 x_1 - n_1 x_2)\sqrt{\frac{(n_1 + n_2)}{n_1 n_2 (x_1 + x_2)(n_1 + n_2 - x_1 - x_2)}} \qquad (36.3)$$

Critical region, inference, P-value: Theory has established that z_T in (36.2) is a close approximation to a standard normal z-variate when H_T is true. Here, the two-tailed critical region is therefore from $-\infty$ to $-z_{\alpha/2}$ and from $z_{\alpha/2}$ to $+\infty$. Equivalently, we reject H_T in favor of H_A whenever $|z_T| > z_{\alpha/2}$. The exceedance probability is $P = 2P(z > |z_T|)$.

Confidence intervals: If H_T is accepted, the $100(1 - \alpha)\%$ confidence interval for $(\pi_1 - \pi_2)$ can be calculated as

$$(p_1 - p_2) \mp z_{\alpha/2}\sqrt{\frac{pq(n_1 + n_2)}{n_1 n_2}} \qquad (36.4)$$

with $p = (x_1 + x_2)/(n_1 + n_2)$. Preferably, and for sure if H_A is accepted, allowance is made for differing variances by calculating the interval as

$$(p_1 - p_2) \mp z_{\alpha/2}\sqrt{\frac{p_1 q_1}{n_1} + \frac{p_2 q_2}{n_2}} \qquad (36.5)$$

Example: In 1962, L. R. Fisher and Zena D. Hosking reported on the vitamin content of herring: "... vitamin A was found in ... 110 out of 367 testes and in 201 out of 390 ovaries. The frequency with which it was found in the female was significantly different from that in the male, $P < 0.01$." Let us re-examine this finding.

Hypothesis specification: A two-sided H_A is suggested. With f and m for female and male, we specify:

$$H_T:\ \pi_f = \pi_m, \quad H_A:\ \pi_f \neq \pi_m; \qquad 0.01$$

Data:

Sex	x	n	p
f	201	390	0.515
m	110	367	0.300
	311	757	0.411

Assumptions: The numbers 390 and 367 are large enough to ensure that, sufficiently closely, $p_f \sim N(\pi_f, \pi_f Q_f / 390)$ and $p_m \sim N(\pi_m, \pi_m Q_m / 367)$.

Test statistic: If H_T is true we estimate π as $p = 311/757 = 0.411$ so that, (36.2) gives

$$z_T = \frac{(0.515 - 0.300)}{\sqrt{(0.411)(0.589)(757)/(390)(367)}} = 6.0$$

or, from (36.3),

$$z_T = \{367(201) - 390(110)\} \sqrt{\frac{757}{390(367)(311)(446)}}$$
$$= 6.0$$

Critical value, inference, exceedance probability: For a two-sided H_A with $\alpha = 0.01$, $z_{0.005} = 2.576$ which is exceeded by 6.0, the test statistic. We can confidently accept H_A.

The exceedance probability $P(z > 6.0)$ is too small to be obtained from the usual tabulations. There seems little risk of a Type I error in accepting H_A here.

Confidence interval: From (36.5) the 99% confidence interval for $\pi_f - \pi_m$ is

$$(0.515 - 0.300) \mp 2.576 \sqrt{\frac{0.515(0.485)}{390} + \frac{0.300(0.700)}{367}}$$
$$= 0.125, 0.305$$

§37. χ^2-TESTS FOR PROPORTIONS

A z^2-variate is a $\chi^2(1)$-variate and hence χ^2 procedures can be used for testing proportions in the §35, §36, and other situations. In all cases, the test statistic is basically

$$\chi_T^2 = \sum_{\substack{\text{all} \\ \text{classes}}} \frac{(O_i - H_i)^2}{H_i} \tag{37.1}$$

where O_i is the number actually observed in the ith class, and H_i is the number predicted for the ith class on the assumption that test hypothesis H_T is true.

Notes: (i) It is essential to make the (37.1) summation over *all the classes*—one for the successes and one for the failures, that is, two—when testing a single proportion, for example, otherwise the test statistics cannot validly be tested against critical values from χ^2 reference distributions.

(ii) Because of the squaring in the numerator of (37.1), the χ^2 procedure is most convenient in two-sided H_A situations. In, but only in, one degree of freedom cases, however, the square root of the test statistic (37.1) is exactly equal to z_T so that, appropriately signed, $z_T = \sqrt{\chi_T^2}$ can be used as the test statistic when H_A is one-sided.

(iii) If theoretical proportions are specified in H_T, they are used to obtain the H_i-values; in other cases the H_i-values are calculated from the data themselves. H_i-values are not rounded to whole numbers.

§38. TESTING THE AGREEMENT BETWEEN AN OBSERVED AND A THEORETICAL PROPORTION (COMPARE §35, 2)

Hypothesis specification: $H_T: \pi = \pi_T,$ $H_A: \pi \neq \pi_T;$ $\alpha.$

Data: x "successes" are observed in n "trials."

Assumption: n and π are such that $x \sim N(n\pi, n\pi Q)$, with $Q = 1 - \pi$.

Test statistic: Even though the data give only the number of successes, the failure class must also be introduced; there are two classes therefore. The observed numbers are $O_1 = x$ for the success class *and* $O_2 = (n - x)$ for the failure class. The numbers predicted for the two classes, presuming that H_T is true, are the population mean numbers of successes and failures which, from (32.1), are $H_1 = n\pi_T$ and $H_2 = nQ_T = n(1 - \pi_T)$. The two deviations to be squared for (37.1) are $(O_1 - H_1) = (x - n\pi_T)$ and $(O_2 - H_2) = (n - x - nQ_T) = (n - x) - n(1 - \pi_T) = -(x - n\pi_T)$. The test statistic in the form (37.1) is then

$$\chi_T^2 = \frac{(x - n\pi_T)^2}{n\pi_T} + \frac{(n - x - nQ_T)^2}{nQ_T} \tag{38.1}$$

which, because

$$\frac{(x - n\pi_T)^2}{n\pi_T} + \frac{(x - n\pi_T)^2}{nQ_T} = (x - n\pi_T)^2 \left(\frac{1}{n\pi_T} + \frac{1}{nQ_T} \right)$$

$$= \frac{(x - n\pi_T)^2}{n\pi_T Q_T}$$

is just the square of z_T in (35.6).

Critical region, inference, P-value: Because the two $(O_i - H_i)$ deviations sum to zero, they are not independent; that is, one determines the other. Correspondingly, the critical region is obtained from the right-tail of that χ^2-distribution with not two, but one degree of freedom. We accordingly accept H_A if $\chi_T^2 > \chi^2(1; \alpha)$ and H_T otherwise. The exceedance probability is $P[\chi^2(1) > \chi_T^2]$.

Confidence intervals: $100(1 - \alpha)\%$ confidence intervals for the population mean proportion of successes are calculated exactly as in (35.7).

Example: Based on the data in the §35 example, let us examine the question "Was the population proportion of those favoring the land-use bill different from 1/2?", so that a two-sided H_A is now appropriate.

Hypothesis specification: H_T: $\pi = 1/2$, H_A: $\pi \neq 1/2$; $\quad \alpha = 0.05$.

Data: We have 140 "successes," and so 100 "failures," in 240 "trials."

Test statistic: If $\pi = 1/2$ we predict $H_T = 240(1/2) = 120$ successes and also, in this case, 120 failures, giving the table

	Success	Failure	Row Total
Observed (O)	140	100	240
Predicted (H)	120	120	240
$O-H$	20	-20	0

in which it is to be noted that $\sum (O_i - H_i)$ must be zero because $\sum O_i = \sum H_i = n$. The test statistic is then, from (38.1),

$$\chi_T^2 = \frac{20^2}{120} + \frac{(-20)^2}{120} = 6.67$$

This exceeds $\chi^2(1; 0.05) = 3.841$ so that the alternative hypothesis is indicated.

P- Value: The exceedance probability is $P[\chi^2(1) > 6.67]$ which, from the $\chi^2(1)$ distribution table, is slightly less than 0.01.

Confidence interval: The 95% confidence interval for π is $(0.52, 0.65)$ from (35.7).

§39. TESTING THE AGREEMENT BETWEEN OBSERVED AND THEORETICAL PROPORTIONS IN TWO OR MORE CLASSES

The §38 procedure can be extended to situations when there are $k \geq 2$ outcome classes; for example, six when a die is thrown.

Hypothesis specification: If π_i is the population proportion and π_{Ti} is the proportion theoretically obtaining for the ith class, the specification is

$$H_T: \pi_i = \pi_{Ti}, \quad H_A: \text{not } H_T; \quad \alpha$$

$$\text{for } i = 1, \ldots, k \quad \text{and} \quad \sum_i \pi_i = \sum \pi_{Ti} = 1$$

Data: x_i outcomes are observed for the ith class, $i = 1, \ldots, k$, and $N = \sum x_i$ is the total number of outcomes observed.

Assumption: The assumption is that the x_i's are "k-nomially" distributed with N large enough so that the test statistic is, to a close approximation, a $\chi^2(k-1)$-variate when H_T is true. On this, a long-standing empirical recommendation is that no H_i should be less than 5 but a more recent specific recommendation (Yarnold, 1970; see also Roscoe and Byars, 1971) is that the minimum H_i can be as small as

$$\left(\frac{5}{k}\right) \text{ (the number of classes with } H_i < 5) \tag{39.1}$$

Test statistic: The observed number for the ith class is $O_i = x_i$; the predicted number for the ith class is calculated as $H_i = N\pi_{Ti}$ and the test statistic is

$$\chi_T^2 = \sum_{i=1}^{k} \frac{(O_i - H_i)^2}{H_i} \tag{39.2}$$

Critical region, inference, P-value: Because $\sum H_i = \sum O_i = N$, $\sum(O_i - H_i) = 0$ so that any set of $(k-1)$ deviations automatically determines the remaining one. The critical region is obtained accordingly from the

$\chi^2(k-1)$-distribution; we reject H_T in favor of H_A if $\chi_T^2 > \chi^2(k-1; \alpha)$. The exceedance probability is $P = P[\chi^2(k-1) > \chi_T^2]$.

Example: The progeny in a genetics experiment were expected to segregate into classes A, B, C, and D in the ratios $9:3:3:1$. The numbers actually observed in the respective classes were 267, 86, 122, and 25. Were these data statistically compatible with the theoretically expected numbers?

Hypothesis specification: Taking $\alpha = 0.05$, for example, we can write

$$H_T: \pi_A = \frac{9}{16}, \quad \pi_B = \pi_C = \frac{3}{16}, \quad \pi_D = \frac{1}{16}, \quad H_A: \text{not } H_T; \quad \alpha = 0.05$$

Data: $O_A = 267$, $O_B = 86$, $O_C = 122$, $O_D = 25$.

Assumption: All the predicted numbers appreciably exceed 5 for this data. The assumption appears quite safe.

Test statistic: The total number of observations is $267 + 86 + 122 + 25 = 500$. Hence, $H_A = 9(500)/16$, $H_B = H_C = 3(500)/16$, and $H_D = 500/16$, giving the table

	A	B	C	D	Total
O_i	267	86	122	25	500
H_i	281.25	93.75	93.75	31.25	500
$O_i - H_i$	-14.25	-7.75	28.25	-6.25	0

The test statistic is

$$\chi_T^2 = \frac{(14.25)^2}{281.25} + \frac{(7.75)^2}{93.75} + \frac{(28.25)^2}{93.75} + \frac{(6.25)^2}{31.25} = 11.093$$

Critical value, inference P-value: From the χ^2 distribution table for $k - 1 = 3$ degrees of freedom, $\chi^2(3; 0.05) = 7.815$ and, since $11.093 > 7.815$, the alternative hypothesis is indicated, perhaps as a result of the excessive number of progeny observed in class C. The P-value, $P[\chi^2(3) > 11.093]$, slightly exceeds 0.01.

§40. COMPARING TWO INDEPENDENT GROUP PROPORTIONS

The χ^2-test procedure is applied here as an alternative, equivalent analysis for the situation in §36, where estimates $\hat{\pi}_1 = p_1 = x_1/n_1$ and $\hat{\pi}_2 = p_2 = x_2/n_2$ were used to compare two population proportions π_1 and π_2. The hypothesis specification, H_T: $\pi_1 = \pi_2$, H_A: $\pi_1 \neq \pi_2$: α, and the data are exactly as in §36.

Assumption: As shown below, the χ_T^2 summation is made for four classes and, for the safety of the normal distribution approximation, the Yarnold (1970) criterion suggests that the predicted number for one, but only one, of the classes may be as low as 1.25.

Test statistic: Recognizing success and failure classes for each group, there are $2 \times 2 = 4$ classes. Identifying row and column by subscripts i and j, respectively, we shall need, for each class, the observed number O_{ij} and the predicted, test-hypothesized, number H_{ij}, as arranged in Table 40.1.

To obtain the H_{ij} we presume that H_T is true so that $\pi_1 = \pi_2 = \pi$ say, in which case the best estimate of the common proportion of successes is the combined number of successes $(x_1 + x_2)$ divided by the total number of opportunities, $n_1 + n_2 = N$ say, to give $\hat{\pi} = (x_1 + x_2)/N$. Out of the n_1 and n_2 opportunities, for groups 1 and 2, respectively, we can accordingly predict the number of successes as $n_1\hat{\pi} = n_1(x_1 + x_2)/N = H_{11}$ and $n_2\hat{\pi} = n_2(x_1 + x_2)/N = H_{12}$. The corresponding predicted numbers of

Table 40.1. The 2×2 Data Table for Comparing Two Group Proportions

	Group	1	2	Row Total
	O	$O_{11} = x_1$	$O_{12} = x_2$	$R_1 = x_1 + x_2$
Success	H	$H_{11} = n_1(x_1 + x_2)/N$	$H_{12} = n_2(x_1 + x_2)/N$	R_1
	$O - H$	$O_{11} - H_{11}$	$O_{12} - H_{12}$	0
	O	$O_{21} = n_1 - x_1$	$O_{22} = n_2 - x_2$	$R_2 = N - R_1$
Failure	H	$H_{21} = n_1 - H_{11}$	$H_{22} = n_2 - H_{12}$	R_2
	$O - H$	$O_{21} - H_{21}$	$O_{22} - H_{22}$	0
Total (O) = total (H)		$C_1 = n_1$	$C_2 = n_2$	$N = n_1 + n_2$
Total $(O - H)$		0	0	0

failures are then immediately obtained by the subtractions $H_{21} = n_1 - H_{11}$ and $H_{22} = n_2 - H_{22}$. Three facts are conveniently noted at this stage:

(i) H_{11} is readily computed as the product

$$H_{11} = \frac{R_1 C_1}{N} \qquad (40.1)$$

where $R_1 = x_1 + x_2$ is the total of the observed numbers in the first row and $C_1 = n_1$ is the total of the numbers observed in the first column. A corresponding result holds for each of the H_{ij}.

(ii) It is really only necessary to compute H_{11} directly, because

$$H_{12} = \frac{R_1 C_2}{N} = \frac{R_1 (N - C_1)}{N} = R_1 - \frac{R_1 C_1}{N} = R_1 - H_{11}$$

(iii) In practical cases the deviation $(O_{11} - H_{11})$ is obtained by subtraction of the numerical values. Its algebraic expression is easily shown to be

$$O_{11} - H_{11} = \frac{n_2 x_1 - n_1 x_2}{N} = \frac{O_{11} O_{22} - O_{21} O_{12}}{N} \qquad (40.2)$$

Also shown in Table 40.1 are the facts that the $(O_{ij} - H_{ij})$ deviations add to zero by both rows and columns so that

$$(O_{12} - H_{12}) = (O_{21} - H_{21}) = -(O_{11} - H_{11})$$

The test statistic can then be calculated in accordance with (37.1) as

$$\chi_T^2 = (O_{11} - H_{11})^2 \left(\frac{1}{H_{11}} + \frac{1}{H_{12}} + \frac{1}{H_{21}} + \frac{1}{H_{22}} \right) \qquad (40.3)$$

Alternatively, exactly the same χ_T^2 can be calculated from the algebraically equivalent expression:

$$\frac{N(O_{11} O_{22} - O_{21} O_{12})^2}{R_1 R_2 C_1 C_2} \qquad (40.4)$$

Notes: (i) The H_{ij} should not be rounded to integer numbers. (ii) χ_T^2 in (40.3) is the square of z_T in (36.3). (iii) It is not an uncommon oversight, especially if only the proportions of successes are given, to calculate a test statistic like that in (40.3) using only two classes. This is erroneous. (iv) It is also not uncommon, and also erroneous, to compute the test statistic using proportions or percentages instead of first converting these into the corresponding observed numbers O_{ij}.

Critical region, inference, P-value: All but one of the $(O_{ij} - H_{ij})$ deviations can be obtained by subtractions. Essentially, only one of the $(O_{ij} - H_{ij})^2$ is involved in the test statistic and the critical region is accordingly obtained from the $\chi^2(1)$-distribution. The two-sided alternative hypothesis is accepted if $\chi_T^2 > \chi^2(1; \alpha)$ with a P-value $P = P[\chi^2(1) > \chi_T^2]$. Otherwise, we stay with H_T. If H_A is one-sided we can give $z_T = \sqrt{\chi_T^2}$ the appropriate sign and use a one-tailed critical region from the z-distribution.

$100(1 - \alpha)\%$ Confidence intervals: For π_1 we substitute p_1 and n_1 in (35.7). For $(\pi_1 - \pi_2)$ we use (36.5) and if it is decisively inferred that $\pi_1 = \pi_2 = \pi$ we can substitute $p = (x_1 + x_2)/N$ and $n = N$ in (35.7).

Example: Using the same data as in §36, Table 40.1 gives Table 40.2, where, for example, from (40.1), $H_{11} = (311)(390)/757 = 160.2$. The test statistic can be calculated from either (40.3) or (40.4), respectively, as

$$\chi_T^2 = (40.8)^2 \left(\frac{1}{160.2} + \frac{1}{150.8} + \frac{1}{229.8} + \frac{1}{216.2} \right) = 36.4$$

$$\chi_T^2 = \frac{757\{201(257) - 189(110)\}^2}{(311)(446)(390)(367)} = 36.4$$

This decisively exceeds the value $\chi^2(1; 0.01) = 6.635$ and we infer that the two population proportions are not the same. The analysis is completed exactly as in §36.

Table 40.2. The 2×2 Arrangement for Examining Two Proportions

	Group	f	m	Row Total
Vitamin A	O	201	110	311
	H	160.2	150.8	311
	$O - H$	40.8	−40.8	0
No vitamin A	O	189	257	446
	H	229.8	216.2	446
	$O - H$	−40.8	40.8	0
Total O = total H		390	367	757

§41. TESTING 2 × 2 CONTINGENCY TABLE DATA FOR ASSOCIATION BETWEEN THE TWO FACTORS

Investigation context: Suppose that each of N individuals gives correct answers to the two questions: (i) Do you have poor or good hearing ability? and (ii) Are you subject to light or heavy noise exposure? Each individual can then be classified into one of the $2 \times 2 = 4$ cells of what is termed a 2×2 contingency table. The cell counts of such tables can then be analyzed to address the question: Are the two binomially classified factors or attributes, hearing ability and noise exposure in the example, independent or associated? The arithmetic of the analysis is exactly that used in §40. The general procedure is now illustrated using the following example.

Example: Respecting the myth that students fail their next test if they walk on the ISU zodiac signs, *The Des Moines Register* (28 Feb. 1971) reported that—of 458 Iowa State University students, 195 walked on the zodiac and 263 skirted round it. Of the 195 who walked on it, 60 were female and 135 were male. Of the 263 that avoided the sign, 110 were females.

Hypothesis specification: Accepting a Type I error probability α, we write

> H_T: the attributes are independent, H_A: the attributes are associated; the two attributes in the example being (i) zodiac sign walking behavior and (ii) sex.

Data: N individuals are classified into one or the other outcome class for each attribute. The numbers of individuals O_{ij}, in the ith row cell, for $i = 1$ and 2, and the jth column cell, for $j = 1$ and 2, are recorded as in Table 41 where $O_{12} = 135$ for the example.

Assumptions: N is large enough to support the normal distribution approximation. With such a large $N = 458$, all the predicted H_{ij}-values are also comfortably large and the assumption is safe.

Test statistic: The test statistic is

$$\chi_T^2 = N^{-2}(O_{11}O_{22} - O_{12}O_{21})^2 \left(\frac{1}{H_{11}} + \frac{1}{H_{12}} + \frac{1}{H_{21}} + \frac{1}{H_{22}}\right) \qquad (41.1)$$

which, with H_{ij} = number predicted for the i,jth cell presuming H_T is

Table 41. Peregrination and Sex Contingency Table Data

			Second Attribute		
			f	m	Row Total
	Walked on	O	60	135	195
	Z-signs	H	72.38	122.62	195
		$O-H$	−12.38	12.38	0
First attribute	Skirted	O	110	153	263
	Z-signs	H	97.62	165.38	263
		$O-H$	12.38	−12.38	0
Column total (O, H)			170	288	458

true, is algebraically equivalent to (40.3) and (40.4). The slightly different argument leading to the H_{ij}-values runs as follows. Suppose that the two attributes *are* independent; the estimate of the probability π_1 that a random individual will walk on the signs is R_1/N, where R_1 is the first row total. Similarly, the estimate of the probability π_2 that a randomly selected individual is male, is estimated by C_2/N, where C_2 is the second column total. Hence, by the axiom (30.1), the probability that a random individual walks on the signs *and* is male is $\pi_1\pi_2$, estimated as R_1C_2/N^2, so that the number predicted for the 1, 2th cell is $H_{12} = N(R_1C_2/N^2) = R_1C_2/N$, as in §40. The remaining H_{ij} can be similarly calculated or found by subtraction from the appropriate marginal totals.

For the example, (41.1) gives

$$\chi_T^2 = \left(\frac{1}{458}\right)^2 \{60(153)-110(135)\}^2 \left(\frac{1}{72.38}+\frac{1}{122.62}+\frac{1}{97.62}+\frac{1}{165.38}\right)$$
$$= 5.85$$

or, from (40.4),

$$\chi_T^2 = \frac{N(O_{11}O_{22}-O_{12}O_{21})^2}{R_1R_2C_1C_2} = \frac{458\{60(153)-110(135)\}^2}{(195)(263)(170)(288)} = 5.86$$

Inference: The critical region again consists of the right-tail $\chi^2(1)$-variate values. We reject H_T and infer association between the two attributes if $\chi_T^2 > \chi^2(1; \alpha)$ or independence of the attributes if $\chi_T^2 < \chi^2(1; \alpha)$. For the exceedance probability, Table A4 shows that we can write $0.01 < P < 0.025$ and because $\chi_T^2 > \chi^2(1; 0.025) = 5.024,$ we infer association between the two attributes under examination and may speculate about its source.

Measuring the association: It is sometimes interesting to measure the "strength of the association" on a scale from 0, for independence, to 1, for complete association. The *coefficient of association*, $\sqrt{\chi_T^2/N}$, does this for the above investigation context. The coefficient is analogous to the correlation coefficient between two continuous variates to be examined later. The value $\sqrt{5.86/458} = 0.113$ is small for the example, but the above analysis can be interpreted as showing that 0.113 is significantly different from zero as are small correlation coefficients obtained from very large samples.

§42. TESTING FOR ASSOCIATION BETWEEN TWO FACTORS WITH TWO OR MORE OUTCOMES: THE $r \times c$ CONTINGENCY TABLE

Generalizations of the previous situation occur when each of N individuals is classified into one of $r \geq 2$ classes for the first factor and also into one of $c \geq 2$ classes for the second factor. The test and alternative hypotheses are again H_T: the two factors are independent, H_A: the two factors are associated. The observed and test-hypothesized numbers, O_{ij} and H_{ij}, are tabulated for each of the $(r)(c)$ classes where H_{ij}, the number predicted for the cell in the ith of the r rows and the jth of the c columns presuming that H_T is true, is found, by the argument used in §41, to be again

$$H_{ij} = \frac{(i\text{th row total})(j\text{th column total})}{N} \qquad (42.1)$$

The test statistic can be calculated, as before, as

$$\chi_T^2 = \sum_i \sum_j \frac{(O_{ij} - H_{ij})^2}{H_{ij}} \qquad (42.2)$$

or by equivalent form, which avoids the differencing operation,

$$\chi_T^2 = \left(\sum_i \sum_j \frac{O_{ij}^2}{H_{ij}} \right) - N \qquad (42.3)$$

or, but only when $r = 2$ (Cox, 1982; Good, 1983), with half the number of summations, as

$$\chi_T^2 = N - \frac{N^2}{R_1 R_2} \sum_j \frac{O_{1j} O_{2j}}{C_j} \qquad (42.4)$$

where R_1 and R_2 are the two row totals and C_j is the jth column total.

Only $(r-1)(c-1)$ of the $(O_{ij} - H_{ij})$'s can be independent, the remaining ones being obtained by subtractions from the appropriate marginal totals. Assuming, for the normality approximation, that N is large enough, which may be checked using the criterion in (39.1), the critical region is obtained from the χ^2-distribution with $(r-1)(c-1)$ degrees of freedom so that H_T is rejected in favor of H_A if $\chi_T^2 > \chi^2\{(r-1)(c-1); \alpha\}$.

Example: Table 42.1 shows the observed and hypothesized numbers when the resting zone heights, $c = 4$, were observed in $r = 3$ seasons for $N = 1123$ flies.

The H_{ij} are computed from (42.1), for example, $H_{12} = (218)(437)/1123$ and $H_{34} = (340)(88)/1123$, and entered beneath the O_{ij}, so that the test statistic, using (42.3), is obtained as

$$\chi_T^2 = -1123 + \left\{ \frac{(24)^2}{72.60} + \frac{(120)^2}{84.83} + \cdots + \frac{(31)^2}{26.64} \right\} = 150.81$$

For $(r-1)(c-1) = 6$ degrees of freedom, χ_T^2 greatly exceeds $\chi^2(6; 0.005) = 18.548$ so that H_A is indicated; that is, resting height did depend on, was associated with, season.

More advanced procedures, such as log-linear model analysis, are available in Fienberg (1977), for example, for exploring the nature of

Table 42.1. Flies Classified by Resting Height and Season

		Height Zone (feet)				
		0–3	3–6	6–9	>9	Total
Spring	O	24	120	57	17	218
	H	72.60	84.83	43.48	17.08	218
Summer	O	282	164	79	40	565
	H	188.17	219.86	112.70	44.27	565
Autumn	O	68	153	88	31	340
	H	113.23	132.31	67.82	26.64	340
Total	(O, H)	374	437	224	88	1123

Table 42.2. A 2×4 Contingency Table of Fly Counts

	Height Zone (feet)				
	0–3	3–6	6–9	>9	Total
Spring	24	120	57	17	218
Autumn	68	153	88	31	340
Total	92	273	145	48	558

such dependence. Here, noting the meteorological similarities of Spring and Autumn—and their differences from the Summer scene—we might ask whether association is revealed if the Summer counts are omitted to give the data in Table 42.2, a 2×4 contingency table. The test statistic can be computed as before using (42.2) or (42.3), or we can now use (42.4) to avoid the direct calculation of the individual H_{ij}. We find

$$\chi_T^2 = 558 - \frac{558^2}{340(218)} \left\{ \frac{24(68)}{92} + \frac{120(153)}{273} + \frac{57(88)}{145} + \frac{17(31)}{48} \right\} = 9.54$$

Then, using right-tail critical regions from the χ^2-distribution with $(2-1)(4-1) = 3$ degrees of freedom, we find $11.345 > 9.54 > 9.348$ so that, for the exceedance probability, $0.01 < P < 0.025$. There is still evidence that resting height and season are associated.

§43. χ^2-TESTS FOR DISTRIBUTION ASSUMPTIONS

A χ^2-test procedure is occasionally useful for examining the assumption that a set of data can reasonably be regarded as a sample from some specified distribution. For this, the observations are counted into variate-value classes in a frequency distribution, O_i being the number in the ith class. The population parameters for the specified distribution are taken to be equal to their sample estimates which are then used to generate a predicted or hypothesized number H_i for each class. The test statistic, $\chi_T^2 = \sum (O_i - H_i)^2 / H_i$ as in (37.1), is then tested against a $\chi^2(f)$ reference distribution, where the number of degrees of freedom, f, is

$f =$ (the number of classes) − (the number of parameters estimated) − 1

$$\tag{43.1}$$

No matter what specified distribution is being tested as the "parent"

Table 43. A Sample Frequency Distribution of 200 Goat Gestation Periods (days)

Period	<145.5	146	147	148	149	150	151	152	153	154	>154.5
O_i	7	4	13	20	28	44	31	18	22	6	7
H_i	6.8	8.4	15.0	22.2	28.6	31.8	29.4	23.8	16.2	9.8	8.0

distribution for the sample, it has been theoretically established that the test statistic is, approximately, a $\chi^2(f)$-variate *provided the sample size is large enough*. The requisite assumption that the sample size is large enough may be assessed using the criterion (39.1) and, if necessary, several contiguous classes can be coalesced into just one to avoid H_i-numbers that are too small to meet the criterion.

Example: In Table 43, except for the lower and upper intervals, the O_i-values show the numbers of goats with gestation periods falling in the intervals from -0.5 to $+0.5$ days of the day shown in the first row. The original daily classes at the ends of the distribution have been combined to avoid small H_i-values.

The data will be used to examine the hypothesis that the gestation period variate, x say, is normally distributed. Estimates $\hat{\mu} = 150.1$ and $\hat{\sigma} = 2.514$ were first calculated from the original data and, presuming that $x \sim N(150.1, 2.514^2)$, the H_i-value for each class was obtained as $H_i = $ (total number in sample) π_i, where π_i is the probability that a random member of the specified normal distribution will fall into the ith class. For example, as in §11, Example (i, b), we find for the 150-day class,

$$\pi_6 = P[149.5 < x < 150.5] = P[-0.239 < z < 0.159] = 0.159$$

so that $H_6 = 200(0.159) = 31.8$. After, advisedly, checking that $\sum H_i = 200$, the test statistic is computed as

$$\chi_T^2 = \sum \frac{(O_i - H_i)^2}{H_i} = \frac{(0.2)^2}{6.8} + \frac{(4.4)^2}{8.4} + \cdots + \frac{(1.0)^2}{8.0} = 12.67$$

With 11 classes and 2 parameters, μ and σ, estimated, (43.1) gives $f = 11 - 2 - 1 = 8$; the $\chi^2(8)$ reference distribution shows that $\chi^2(8; 0.05) = 15.507$ which exceeds $\chi_T^2 = 12.67$ so that, at $\alpha = 0.05$, the test does not reject the hypothesis that x is normally distributed. The test, of course, does not establish that x *is* necessarily normally distributed and some other symmetrical distribution could well be superior, perhaps by reducing the relatively large deviation, $(O_i - H_i) = 12.2$, for the 150-day class.

The literature of testing for non-normality is understandably extensive; for example, Martinez and Iglewicz (1981) have proposed a test when the actual values, instead of a frequency distribution, are available for the variate; see also Iman (1982) and Mage (1982).

§44. THE POISSON DISTRIBUTION

If in a binomial distribution situation, π, the probability of success per trial, tends to zero while n, the number of trials, tends to infinity in such a way that $n\pi$, the binomial mean number of successes, stays constant, $n\pi = \mu$ say, then in the limit, x, the number-of-successes-variate, has a Poisson distribution. The probability that exactly x successes will be observed is given by

$$P(x) = \frac{\mu^x e^{-\mu}}{x!}, \qquad x = 0, 1, 2, \ldots \tag{44.1}$$

where $e = 2.71828 \cdots$ is the base of natural logarithms.

Characteristics of a Poisson variate:

(i) In theory, the Poisson variate x can take any value from zero to infinity.

(ii) The probabilities of obtaining exactly $(x + 1)$ and exactly x successes are related, from (44.1), as

$$\frac{P(x + 1)}{P(x)} = \frac{\mu}{x + 1} \tag{44.2}$$

(iii) A Poisson distribution is a one-parameter distribution because the probability (44.1) can be found for any specified x-value if the parameter μ is known.

(iv) The best sample estimate of μ is the arithmetic mean number of successes.

(v) The variance of the distribution is equal to the mean, that is,

$$V(x) = \mu[(x - \mu)^2] = \mu \tag{44.3}$$

(vi) Independent Poisson variates are additive. Thus, if x_1 is a Poisson variate with mean and variance μ_1 and if x_2 is another Poisson variate with parameter μ_2, then $(x_1 + x_2)$ is a Poisson variate with mean and variance $(\mu_1 + \mu_2)$, and so on.

§45. IS THIS VARIATE NOT POISSON DISTRIBUTED? A χ^2-TEST

The distribution testing procedure in §43 is illustrated here with the bacteriophage data from Ellis and Delbruck (1939) in Table 45.1. To examine the hypothesis that x has a Poisson distribution, we first estimate μ from (iv) above as

$$\hat{\mu} = \frac{14(1) + 5(2) + 1(3)}{(13 + 14 + 5 + 1)} = 0.8182$$

Now assuming that x has a Poisson distribution with parameter $\mu = 0.8182$, we use (44.1) to give the probability that $x = 0$ as $e^{-\mu}$ (remembering that $0! = 1$) so that out of 33 plates, the number of plates showing zero plaques will be, for the first class,

$$H_1 = \frac{33}{(2.71828)^{0.8182}} = 14.561$$

Then using (44.2) in succession,

$$H_2 = \frac{0.8182 H_1}{1} = (0.8182)(14.561) = 11.914$$

$$H_3 = \frac{0.8182 H_2}{2} = (0.4141)(11.914) = 4.874$$

Finally, to complete Table 45.2, we combine the remaining classes to avoid small H_i-numbers and find the hypothesized number of plates showing three or more plaques as

$$33 - 14.561 - 11.914 - 4.874 = 1.651$$

Table 45.1. Frequency Distribution for x, the Number of Plaques

x	0	1	2	3	≥ 4
Number of plates	13	14	5	1	0

Table 45.2. O_i and H_i Numbers for Testing a Possibly Poisson Distributed Variate

x	0	1	2	≥ 3	Total
O_i	13	14	5	1	33
E_i	14.561	11.914	4.874	1.651	33
$O_i - E_i$	−1.561	2.086	0.126	−0.651	0

The test statistic (37.1) is

$$\chi_T^2 = \frac{(1.561)^2}{14.561} + \frac{(2.086)^2}{11.914} + \frac{(0.126)^2}{4.874} + \frac{(0.651)^2}{1.651} = 0.79$$

With four classes and one estimated parameter, (43.1) gives $\chi^2(2)$ as the reference distribution and, since 0.79 is considerably less than $\chi^2(2; 0.05) = 5.991$, we find no evidence to reject the assumption that the number of plaques is Poisson distributed.

Linear Regression: Fitting Straight Lines to (x, y) Data

§46. THE STRAIGHT LINE EQUATION

If n pairs of values $(x_1, Y_1), (x_2, Y_2), \ldots, (x_n, Y_n)$ fall exactly on a straight line when plotted on a graph, the relationship between x and Y can be written

$$Y_i = \alpha + \beta x_i \qquad (46.1)$$

which shows that the Y-value corresponding to any x-value can be calculated if the constants α and β, referred to as the parameters of the line, are known. The distinction between this α and the Type I error probability will be clear in context.

Equation (46.1) shows that α, termed the *intercept*, is the Y-value corresponding to $x = 0$, so that α is the distance from the origin to the point where the line meets the Y-axis; see Fig. 46. β is the *slope* of the line, the tangent of the angle between the line and the x-axis. Equivalently, *β is the constant amount by which Y increases when x is increased by one unit.* From (46.1), the Y-values, $Y_{(x)}$ and $Y_{(x+1)}$ say, corresponding to x and $(x + 1)$ are given by

$$Y_{(x)} = \alpha + \beta x \quad \text{and} \quad Y_{(x+1)} = \alpha + \beta(x + 1) = (\alpha + \beta x) + \beta$$

Subtraction gives $Y_{(x+1)} - Y_{(x)} = \beta$.

The units of β and of its estimate b in (48.1) are those of Y per one unit of x.

An alternative form for the line equation is

$$Y_i = \mu + \beta(x_i - \bar{x}) = (\mu - \beta\bar{x}) + \beta x_i \qquad (46.2)$$

91

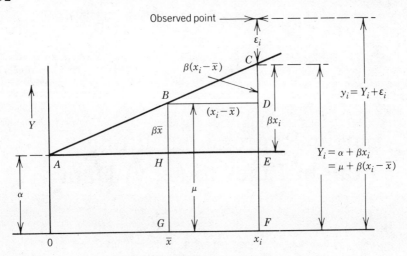

Figure 46. A population regression line.

Here β is again the slope while μ is the Y-value corresponding to $x = \bar{x}$ (Fig. 46). Comparison of (46.1) and (46.2) shows that

$$\alpha = \mu - \beta \bar{x} \qquad (46.3)$$

and so, with $\bar{x}$ known, any two of the three parameters α, β, and μ will completely determine the line. In Fig. 46,

(i) $0A = FE = \alpha$, $AE = x_i$, $EC = \beta x_i$

$\qquad Y_i = FC = FE + EC = \alpha + \beta x_i$

(ii) $BD = GF = (x_i - \bar{x})$, $DC = \beta(x_i - \bar{x})$, $FD = GB = \mu$

$\qquad Y_i = FC = FD + DC = \mu + \beta(x_i - \bar{x})$

(iii) $\alpha = 0A = GH = GB - HB = \mu - \beta \bar{x}$.

§47. LINEAR REGRESSION STATISTICAL MODELS I AND II

In practice, when a sample of n pairs of observations (x_i, y_i), $i = 1, 2, \ldots, n$, is obtained to investigate the relationship, the ordinate y_i at x_i will not be the distance Y_i in (46.1) or (46.2) but will differ from it by a, hopefully small, positive or negative amount, ε_i say, so that the points (x_i, y_i) will not fall exactly on a line. The statistical model for the observations y_i can then be written, using (46.2), for example,

$$y_i = Y_i + \varepsilon_i = \mu + \beta(x_i - \bar{x}) + \varepsilon_i, \quad i = 1, 2, \ldots, n \qquad (47.1)$$

If repeated observations $y_{i1}, y_{i2}, \ldots$ were to be taken at the same x_i-value, they would differ because of differences between the $\varepsilon_{i1}, \varepsilon_{i2}, \ldots$ which are individual members of a conceptual distribution of ε's at this x-value. Taking the population mean of this distribution to be zero, (47.1) shows that $Y_i = \mu(y \,|\, x_i)$ is, reading the "$|$" as "at," the population mean of all the conceptual y_i-values at this x_i. Correspondingly,

$$\mu(y \,|\, x) = \mu + \beta(x - \bar{x}) = \alpha + \beta x \qquad (47.2)$$

is the equation of the underlying *population line*. The statistical problem is to obtain a good estimate of this line by estimating its parameters from $n(x_i, y_i)$ pairs, one y-value being observed at each x-value.

Although the regression analysis procedures are the same for both, it is convenient to recognize two investigation contexts as follows.

Regression model I—fixed x-values: In this, the x-values are fixed, deliberately selected, or specified by the investigator. Crop yields (y) observed at selected fertilizer concentration levels (x) would be an example.

Regression model II—random x-values: Here n individuals or experimental units are taken as a random sample from a bivariate population and both the x and y values are measured for each individual in the sample. As distinct from Model I where one variable has fixed and the other random values, both $x_1, \ldots, x_n$ and $y_1, \ldots, y_n$ are random samples in Model II. Thus, from a random sample of cities, $x =$ city size and $y =$ sulfate concentration could be measured to investigate the rate at which the pollution increases with city size.

Assumptions in regression analysis: The standard linear regression analysis procedures to be described rest on the assumption that the $y_i - \mu(y_i \,|\, x_i) = \varepsilon_i$ quantities in (47.1) are normally and independently distributed with, at each x-value, one and the same population variance $\sigma_{y|x}^2$.

Statistical model for linear regression analysis: Incorporating (47.1) and the foregoing assumptions gives the complete statistical model as

$$y_i = \mu + \beta(x_i - \bar{x}) + \varepsilon_i, \quad \varepsilon_i \sim NI(0, \sigma_{y|x}^2), \quad i = 1, \ldots, n \qquad (47.3)$$

§48. ESTIMATES OF THE REGRESSION PARAMETERS: THE ESTIMATION REGRESSION EQUATION

Based on (47.3) the following can be shown:

(i) The estimate of μ is $\bar{y} = \sum_1^n y_i / n$.

(ii) The estimate of the regression coefficient β is

$$b = \frac{\sum_1^n (x_i - \bar{x})(y_i - \bar{y})}{\sum_1^n (x_i - \bar{x})^2} = \frac{\sum' xy}{\sum' xx} \tag{48.1}$$

(iii) The estimate of the intercept α is $(\bar{y} - b\bar{x})$ as shown in Fig. 48.

(iv) The estimate of $\mu(y \mid x_i)$, the mean of the population of y-values conceptually observable at $x = x_i$, is then $\hat{y}_i$ where

$$\hat{y}_i = \bar{y} + b(x_i - \bar{x}) = (\bar{y} - b\bar{x}) + bx_i \tag{48.2}$$

It may be noted from the equation (48.2) of the estimation or prediction line that it passes through the point $(\bar{x}, \bar{y})$ as in Fig. 48.

The residuals

$$e_i = (y_i - \hat{y}_i) = y_i - \bar{y} - b(x_i - \bar{x}), \qquad i = 1, \ldots, n \tag{48.3}$$

are the differences between the observed values and the estimates, sometimes called the predicted values, of their means, $\mu(y \mid x_i)$. The e_i's

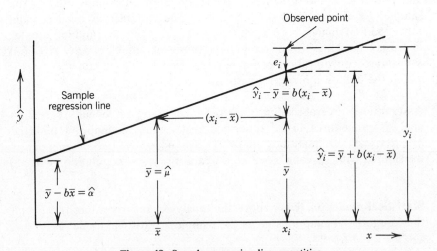

Figure 48. Sample regression line quantities.

will be small if a good fit is achieved and, in fact, the principle of least squares leading to the specific estimates $\bar{y}$ and b above is that these are just the values which make the residual sum of squares $\sum e_i^2 = \sum (y_i - \hat{y}_i)^2$ as small as possible. It can also be shown that the e_i-values satisfy the two conditions:

$$\sum e_i = 0 \quad \text{and} \quad \sum (x_i - \bar{x})e_i = \sum x_i e_i = 0 \qquad (48.4)$$

§49. EXAMPLE: REGRESSION LINE COMPUTATIONS

The form (48.1) can be adapted to a computational form using the fact that

$$\sum' xy = \sum (x_i - \bar{x})(y_i - \bar{y}) = \sum x_i y_i - \frac{(\sum x_i)(\sum y_i)}{n} \qquad (49.1)$$

giving

$$b = \frac{\sum xy - (\sum x)(\sum y)/n}{\sum x^2 - (\sum x)^2/n} = \frac{n \sum xy - (\sum x)(\sum y)}{n \sum x^2 - (\sum x)^2} \qquad (49.2)$$

Example: The computations required in (49.2) are exemplified in Table 49 where it may be noted that the equally spaced and integral x-values suggest that the data refer to a regression Model I situation.

Here $n = 4$ so that $\bar{y} = 20/4 = 5$ and, from (49.2),

$$b = \frac{4(58) - (10)(20)}{4(30) - (100)} = 1.6$$

Since $\bar{x} = 2.5$, the equation of the line of best fit (48.2) is

$$\hat{y} = 5 + 1.6(x - 2.5) = 1.0 + 1.6x$$

Table 49. Regression Line Computations

	x	y	x^2	xy	$y - \hat{y}$
	1	2	1	2	−0.6
	2	5	4	10	0.8
	3	6	9	18	0.2
	4	7	16	28	−0.4
Total	10	20	30	58	0.0

so that the intercept α is estimated as 1.0. Furthermore, at $x = 4$, for example, the predicted value to estimate $\mu(y|4) = \mu + \beta(4 - 2.5)$ is $\hat{y}_4 = 1.0 + (1.6)(4) = 7.4$; the residual is $e_4 = y_4 - \hat{y}_4 = 7.0 - 7.4 = -0.4$. The remaining residuals, similarly calculated, give the values in the last column of Table 49; it may be checked that they satisfy the two conditions (48.4).

§50. THE DISTRIBUTIONS OF THE SAMPLE ESTIMATES $\bar{y}$ AND b

Different repeated sets of n (x, y) pairs, taken in the same regression context, would give different values of $\bar{y}$ and b. The estimates are therefore variates. Thus, we regard the quantity b, calculated from the one particular set of n pairs in hand, as one randomly drawn member of a conceptual population of such b-variates. The calculated value will therefore probably be a good estimate of (i.e., will probably be close to) the population slope β, if the variance of the conceptual population of b-variates is small. To assess this we need the following facts about the distribution of $\bar{y}$ and b.

The variate $\bar{y}$ is normally distributed with population mean μ and population variance $V(\bar{y}) = \sigma^2_{y|x}/n$. We write

$$\bar{y} \sim N(\mu, \sigma^2_{y|x}/n) \tag{50.1}$$

The variate b is normally distributed with population mean $\mu(b) = \beta$ and variance $V(b) = \sigma^2_{y|x}/\sum' xx$. We write

$$b \sim N\left(\beta, \sigma^2_{y|x}\Big/\sum' xx\right) \tag{50.2}$$

The last result has a useful practical implication. It shows that, for best precision, b has the smallest variance when $\sum' xx = \sum (x - \bar{x})^2$ is largest. Hence, if it is *known* that the (x, y) relationship *is linear* in a Model I situation, we should, if n is even, take $n/2$ y-observations at the lower and $n/2$ at the higher end of the x-range being studied. Observations on y's at intermediate x-values will be required if the form of the relationship, "Is it linear or of higher degree in x?," is to be investigated. The procedures in §136 are then appropriate.

Derivations[+]: (May be skipped! See Preface.) The derivations of the preceding results run as follows. For $\bar{y}$, using the model equation (47.3),

$$\bar{y} = \frac{\sum y_i}{n} = \frac{n\mu + \beta \sum (x_i - \bar{x}) + \sum \varepsilon_i}{n}$$

$$= \mu + \bar{\varepsilon} \tag{50.3}$$

since $\sum (x_i - \bar{x}) = 0$. It follows that $\mu(\bar{y}) = \mu$, because $\mu(\bar{\varepsilon}) = 0$, and that $V(\bar{y}) = V(\bar{\varepsilon}) = \sigma^2_{y|x}/n$. Hence, on the assumption that $\varepsilon \sim NI(0, \sigma^2_{y|x})$,

$$\bar{y} \sim N(\mu, \sigma^2_{y|x}/n)$$

For b, its numerator in (48.1) is

$$\sum (x_i - \bar{x})(y_i - \bar{y}) = \sum (x_i - \bar{x}) y_i - \bar{y} \sum (x_i - \bar{x}) = \sum (x_i - \bar{x}) y_i - 0$$

which gives, using (47.3) again,

$$\sum (x_i - \bar{x}) y_i = \sum (x_i - \bar{x})\{\mu + \beta(x_i - \bar{x}) + \varepsilon_i\}$$

$$= 0 + \beta \sum (x_i - \bar{x})^2 + \sum (x_i - \bar{x}) \varepsilon_i$$

Division by $\sum (x_i - \bar{x})^2 = \sum' xx$ shows that

$$b = \beta + \frac{\sum (x_i - \bar{x}) \varepsilon_i}{\sum' xx} \tag{50.4}$$

so that the second term on the right-hand side is the unknown variability quantity, $(b - \beta)$, by which b differs from β. Since the population mean of each ε_i is zero, it follows that the population mean of the variability quantity is zero and hence b is the unbiased estimate of β; that is,

$$\mu(b) = \beta$$

Furthermore,

$$V(b) = V\left\{\sum (x_i - \bar{x}) \varepsilon_i \Big/ \sum' xx\right\} = V\left(\sum c_i \varepsilon_i\right) \tag{50.5}$$

with $c_i = (x_i - \bar{x})/\sum' xx$.

Now extending the rules (6.2) and (7.1) to deal with the sum of n independent variates,

$$V\left(\sum c_i \varepsilon_i\right) = c_1^2 \sigma^2_{y|x} + c_2^2 \sigma^2_{y|x} + \cdots + c_n^2 \sigma^2_{y|x} = \left(\sum c_i^2\right) \sigma^2_{y|x}$$

which, because $\sum c_i^2 = \sum (x_i - \bar{x})^2/(\sum' xx)^2 = 1/\sum' xx$, gives with (50.5)

$$V(b) = \frac{\sigma^2_{y|x}}{\sum' xx}$$

Finally, since like $\bar{y}$, b is a sum of normally distributed variates, it has a normal distribution itself so that altogether, as required,

$$b \sim N\left(\beta, \sigma^2_{y|x}\Big/ \sum' xx\right)$$

§51. ESTIMATING THE RESIDUAL VARIANCE $\sigma^2_{y|x}$

As usual, the variability exhibited by the e_i-deviations is used to estimate the population variance of the ε_i-deviations. Here $e_i = y_i - \hat{\mu}(y_i|x_i) = (y_i - \hat{y}_i)$ and the estimate of $\sigma^2_{y|x}$ is

$$\hat{\sigma}^2_{y|x} = s^2_{y|x} = \frac{\sum e_i^2}{n-2} = \frac{\sum (y_i - \hat{y}_i)^2}{n-2} \tag{51.1}$$

with, now, $(n-2)$ degrees of freedom. The reason for this number of degrees of freedom is that, although there are n individual e_i's, they must be inter-related to satisfy two constraints, those given in (48.4). Any $(n-2)$ of the e_i's therefore necessarily determine the remaining two. Equivalently, the sum of the squares of the n *dependent* e_i's in (51.1) behaves like and is distributed in the pattern of, the sum of the squares of $(n-2)$ *independent* ε-deviations. Hence, since the average $\sum_1^{n-2} \varepsilon^2/(n-2)$ would give an unbiased estimate of $\sigma^2_{y|x}$, in accordance with the definition of population variance in (5.1), the $s^2_{y|x}$ in (51.1) will also be an unbiased estimate, and one we can actually calculate, with the desired property that $\mu(s^2_{y|x}) = \sigma^2_{y|x}$.

The last column of Table 49 gives $s^2_{y|x} = \{0.6^2 + \cdots + 0.4^2\}/2 = 0.6$, 2 df, for the example there, but (51.1) would be less convenient for a more realistic data set. An alternative uses the fact derived below that

$$\sum (y - \hat{y})^2 = \sum' yy - \frac{(\sum' xy)^2}{\sum' xx} \tag{51.2}$$

to give

$$s^2_{y|x} = \frac{1}{n-2} \left\{ \sum' yy - \frac{(\sum' xy)^2}{\sum' xx} \right\} \tag{51.3}$$

wherein alternatives for the second numerator term are

$$\frac{(\sum' xy)^2}{\sum' xx} = b \sum' xy = b^2 \sum' xx \tag{51.4}$$

of which the first is perhaps least susceptible to rounding error. The analysis of variance computations for (51.3) are described and illustrated in the next section.

Derivation[+] of (51.2): Using (48.2) for $\hat{y}_i$, we obtain

$$y_i - \hat{y}_i = y_i - \bar{y} - b(x_i - \bar{x}) = (y_i - \bar{y}) - b(x_i - \bar{x})$$

First squaring both sides and then summing shows that

$$\sum (y_i - \bar{y})^2 = \sum \{(y_i - \bar{y})^2 - 2b(y_i - \bar{y})(x_i - \bar{x}) + b^2(x_i - \bar{x})^2\}$$

$$= \sum' yy - 2b \sum' xy + b^2 \sum' xx \qquad (51.5)$$

The equations (51.4) hold because $b = \sum' xy/\sum' xx$ and replacing the last two terms in (51.5) by their equivalents from (51.4) establishes (51.2).

§52. THE ANALYSIS OF VARIANCE (ANOVA) FOR LINEAR REGRESSION

The relation (51.2) can be written

$$\sum' yy = \sum (y - \bar{y})^2 = \frac{(\sum' xy)^2}{\sum' xx} + \sum (y - \hat{y})^2 \qquad (52.1)$$

showing that the variation measured by $\sum' yy$ is partitioned or "analyzed" into two added parts. The first part $(\sum' xy)^2/\sum' xx$ is known as "the sum of squares due to regression." Literally, (52.1) shows that:

$\sum' yy$, the sum of the squares of the deviations of the
y-observations from the mean $\bar{y}$,

equals

$\left(\sum' xy\right)^2 \Big/ \sum' xx$, the sum of squares due to regression,

plus

$\sum (y - \hat{y})^2$, the sum of squares of the deviations of the
y-observations from the regression line

Furthermore, since $b(x - \bar{x}) = \hat{y} - \bar{y}$, from (48.2), $b^2(x - \bar{x})^2 = (\hat{y} - \bar{y})^2$, so that, for the sum of squares due to regression,

$$\frac{(\sum' xy)^2}{\sum' xx} = b^2 \sum' xx = \sum (\hat{y} - \bar{y})^2$$

and (52.1) can be written

$$\sum (y - \bar{y})^2 = \sum (\hat{y} - \bar{y})^2 + \sum (y - \hat{y})^2 \qquad (52.2)$$

Table 52.1. A Linear Regression ANOVA

Variability Source	df	ss	ms
(b) Regression line	1	$(\sum' xy)^2/\sum' xx$	$(\sum' xy)^2/\sum' xx$
(c) Deviations from regression line	$n-2$	$\sum(y-\hat{y})^2$ (by subtraction)	$s^2_{y\mid x}=\sum(y-\hat{y})^2/(n-2)$
(a) Total	$n-1$	$\sum(y-\bar{y})^2$	$s^2_y=\sum(y-\bar{y}^2)^2/(n-1)$

The ANOVA in Table 52.1 provides a convenient arrangement and computational format for the results (52.1) and (52.2). The abbreviations in the first row are, respectively, df for "degrees of freedom," ss for "sums of squares," and ms for "mean squares."

In Table 52.1 it may be noted that:

(i) The sums of squares are calculated in the order (a), (b), (c), which later is obtained by the subtraction $\sum' yy - (\sum' xy)^2/\sum' xx$ in accordance with (52.1) and (52.2).

(ii) The "df for regression" (1) and for "deviations from regression line" $(n-2)$ add to those for the "total" $(n-1)$.

(iii) The corresponding component sums of squares add to $\sum(y-\bar{y})^2$. The mean squares do not add.

(iv) The mean square in each row is obtained by dividing each sum of squares by its number of degrees of freedom. It is then seen that the mean square for "Deviations from regression line" is just $s^2_{y\mid x}$ as defined in (51.3).

(v) Were it not for the relationship between the y's and the x's, the variability in the y's would be estimated as $s^2_y = \sum(y-\bar{y})^2/(n-1)$. Removing that part of the y-variability, $(\sum' xy)^2/\sum' xx$, which is "due to the regression," reduces the unexplained variability to that estimated by $s^2_{y\mid x}$ and, the stronger the relationship is, the greater the reduction in variance.

The coefficient of determination: This coefficient, r^2, is defined as that fraction of the total ss which is due to or explained by the regression, so that $0 \le r^2 \le 1$, and

$$r^2 = \frac{(\sum' xy)^2}{(\sum' xx)(\sum' yy)} \tag{52.3}$$

provides a measure of the success the regression fitting has achieved in explaining the variability in the y-values.

Example: Cox and Roseberry (1966) reported a heuristic study on the relationship between the variance v_n and the mean sample number $\bar{n}$, in sequential probability ratio tests, using a linear regression analysis of $y = \log v_n$ on $x = \log \bar{n}$ for which part of the data (rounded) are given in Table 52.2.

Table 52.2. Linear Regression Data

						Total	Mean
x	0.86	1.03	1.25	1.74	2.32	7.2	1.44
y	1.38	1.58	2.05	3.01	4.21	12.23	2.45
$y - \hat{y}$	0.08	0.06	0.02	0.03	0.03	0	0

The preliminary computations are

$$\sum\nolimits' xx = \{(0.86)^2 + \cdots + (2.32)^2\} - \frac{(7.2)^2}{5} = 1.405$$

$$\sum\nolimits' xy = \{0.86(1.38) + \cdots + 2.32(4.21)\} - \frac{(7.2)(12.23)}{5} = 2.7701$$

$$\sum\nolimits' yy = \{(1.38)^2 + \cdots + (4.21)^2\} - \frac{(12.23)^2}{5} = 5.4729$$

Hence, from (49.2), $b = 2.7701/1.405 = 1.97$ and the equation (48.2) of the estimation line is

$$\hat{y} = 2.45 + 1.97(x - 1.44) = -0.39 + 1.97x$$

giving the $(y - \hat{y})$ residuals in the last row of Table 52.2.
     The sum of squares due to regression is

$$\frac{(\sum' xy)^2}{\sum' xx} = \frac{(2.7701)^2}{1.405} = 5.4615$$

The residual sum of squares, of deviations, is

$$\sum\nolimits' yy - 5.4615 = 5.4729 - 5.4615 = 0.0114$$

for the ANOVA in Table 52.3.

Table 52.3. A Linear Regression ANOVA

Variability Source	df	ss	ms	F_T
Regression	1	5.4615	5.4615	1437.2
Residual	3	0.0114	$0.0038 = s_{y\|x}^2$	
Total	4	5.4729	$1.3682 = s_y^2$	

Anticipating the ANOVA significance test of H_T: $\beta = 0$, H_A: $\beta \neq 0$, the test statistic in (53.6), $F_T = 5.4615/0.0038 = 1437.2$, is now conveniently calculated. Comparison with $F(1,3; 0.001) = 167.5$ strongly indicates that the population regression coefficient is not zero. Correspondingly, the unexplained variability in the y-values has been greatly reduced, from $s_y^2 = 1.3682$ to $s_{y\|x}^2 = 0.0038$. The coefficient of determination (52.3) is $r^2 = 0.998$.

§53. IS THE SLOPE ZERO? HYPOTHESIS TESTS AND INTERVAL ESTIMATES FOR THE REGRESSION SLOPE PARAMETER

It is commonly required to test the significance (of its difference from zero) of b, the slope estimate. For, if $\beta = 0$, the model (47.3) reduces to $y_i = \mu + \varepsilon_i$ in which case y is unrelated to x and the x-values will neither help to predict population means of y-values nor help to reduce the residual variability. The test follows the principles and procedures in §16 and §17.

Hypothesis specification:

$$H_T: \beta = 0, \quad H_A: \beta \neq 0; \qquad \alpha \qquad (53.1)$$

Test statistic:

$$\frac{b}{\text{estimated standard deviation of } b} \qquad (53.2)$$

For the denominator of (53.2), $V(b) = \sigma_{y|x}^2/\sum' xx$ from (50.2); the estimate of $\sigma_{y|x}^2$ is $s_{y|x}^2$ obtained from the ANOVA, Table 52.1, so that the test statistic is

$$t_T = b\sqrt{\frac{\sum' xx}{s_{y|x}^2}} \qquad (53.3)$$

Critical region and inference: On the assumption that $\varepsilon_i \sim NI(0, \sigma^2_{y|x})$, b itself is a normally distributed variate so that, if H_T: $\beta = 0$ is true, t_T in (53.3) is a t-variate with $(n-2)$ degrees of freedom. Accordingly, H_A will be accepted if $|t_T| > t(n-2; \alpha/2)$ and rejected otherwise.

Confidence interval for β: If b is normally distributed about its population mean β,

$$\frac{b - \beta}{\text{estimated standard deviation of } b}$$

is a $t(n-2)$-variate. Hence, by steps analogous to those leading to (13.5), the limits of the $100(1-\alpha)\%$ confidence interval for β are

$$b \mp t(n-2; \alpha/2) \sqrt{\frac{s^2_{y|x}}{\sum' xx}} \tag{53.4}$$

Example: (Continued from §52)

Data: $b = 1.97$, $\quad s^2_{y|x} = 0.0038$ (3 df), $\quad \sum' xx = 1.405$.

Hypothesis specification: H_T: $\beta = 0$, $\quad H_A$: $\beta \neq 0$; $\qquad 0.01$.

Test statistic: $t_T = 1.97\sqrt{1.405/0.0038} = 37.88$.

Critical region and inference: Because 37.88 is much greater than $t(3; 0.005) = 5.841$, the test statistic has fallen in the critical region. A nonzero slope is strongly indicated. In this investigation, it was also of interest to test H_T: $\beta = 2$; H_A: $\beta \neq 2$ for which $t_T = (1.97 - 2)\sqrt{1.405/0.0038} = -0.58$ and the acceptance of this H_T suggested that the variance v_n was proportional to the square of the mean sample number.

Confidence interval: From (53.4) the limits of the 99% confidence interval for β are

$$1.97 \mp 5.841 \sqrt{\frac{0.0038}{1.405}} = 1.67, 2.27$$

The ANOVA test procedure: The above t-distribution procedures are convenient for one-sided H_A situations and for situations when $\beta_T \neq 0$. For a two-sided test with $\beta_T = 0$, however, the ANOVA provides an attractive alternative procedure. From (53.3)

$$t_T^2 = \frac{b^2 \sum' xx}{s_{y|x}^2} \tag{53.5}$$

In the numerator, from (51.4), $b^2 \sum' xx = (\sum' xy)^2 \sum' xx$, which is the mean square due to the regression line in the ANOVA while $s_{y|x}^2$, in the denominator of (53.5), is the mean square for the residual deviations in the ANOVA. Furthermore, as in the §26 example, it is known that the square of a t-variate with f degrees of freedom is an F-variate with 1 and f degrees of freedom in the numerator and denominator, respectively; that is, $(t_f)^2 \sim F(1, f)$.

Hence, for testing $H_T: \beta = 0$, $H_A: \beta \neq 0$; α, the test statistic can be obtained directly from the ANOVA as

$$F_T = \frac{\text{ms due to regression}}{\text{residual ms}} \tag{53.6}$$

for reference to critical regions from the $F\{1, (n-2)\}$-distribution. Also, for either $H_A: \beta < 0$ or $H_A: \beta > 0$, the absolute magnitude of the t-test statistic can conveniently be obtained as $\sqrt{F_T}$ which, given the appropriate sign, can be compared with critical regions from the $t(n-2)$ tabulation.

Confidence interval for β: Using the facts that $s_{y|x}^2/\sum' xx = b^2(s_{y|x}^2/b^2 \sum' xx)$ and $t^2(n-2; \alpha/2) = F\{1, n-2; \alpha\} = F_\alpha$ say, the $100(1 - \alpha)\%$ confidence interval for β can also be calculated as

$$b\left(1 \mp \sqrt{\frac{F_\alpha}{F_T}}\right) \tag{53.7}$$

Example: For testing $H_T: \beta = 0$, $H_A: \beta \neq 0$ we see from the ANOVA, Table 52.3, that $F_T = 1437.2$, leading to the acceptance of H_A as before. The rounding discrepancy between F_T and t_T^2 is much reduced if the latter is calculated using $b = 1.9716$. The 99% confidence interval for β obtained from (53.7) is the same as before, that is,

$$1.97\left(1 \mp \sqrt{\frac{34.116}{1437.2}}\right) = 1.67, 2.27$$

§54. INTERVAL ESTIMATION FOR THE POPULATION MEAN OF OBSERVATIONS AT ONE SPECIFIED x-VALUE

It is sometimes required to obtain a confidence interval for $\mu(y|x_s)$, the population mean of the distribution of y-values at $x = x_s$, where x_s is a specified x-value which may or may not be one of those, $x_1, \ldots, x_n$, used in the regression fitting. From (48.2) the $\hat{y}$-value at $x = x_s$ is

$$\hat{y}_s = \bar{y} + b(x_s - \bar{x}) \tag{54.1}$$

and this is the point estimate of $\mu(y \mid x_s) = \mu(\hat{y} \mid x_s) = \mu_s$ say. Furthermore

$$V(\hat{y}_s) = V\{\bar{y} + b(x_s - \bar{x})\}$$

and it is a known result that $\bar{y}$ and b are independent variates so that, taking $(x_s - \bar{x})$ as a constant multiplier, Rules 1 and 2 (§6, §7) give

$$V(\hat{y}_s) = V(\bar{y}) + (x_s - \bar{x})^2 V(b)$$

which, with $V(\bar{y})$ and $V(b)$, from (50.1) and (50.2), becomes

$$V(\hat{y}_s) = \sigma_{y|x}^2 \left\{ \frac{1}{n} + \frac{(x_s - \bar{x})^2}{\sum' xx} \right\} \tag{54.2}$$

This variance is estimated as $\hat{V}(\hat{y}_s)$ by replacing the unknown population value $\sigma_{y|x}^2$ by its estimate $s_{y|x}^2$ from the ANOVA.

Hence, on the basic assumption that $\varepsilon_i \sim NI(0, \sigma_{y|x}^2)$, the quantity

$$\frac{\hat{y}_s - \mu_s}{\sqrt{\hat{V}(\hat{y}_s)}} \tag{54.3}$$

is a t-variate with $(n - 2)$ degrees of freedom, those for $s_{y|x}^2$. The argument used to obtain confidence intervals for a population mean (§13) then gives the limits of the $100(1 - \alpha)\%$ confidence interval for μ_s as $\hat{y}_s \mp t_{\alpha/2} \sqrt{\hat{V}(y_s)}$; that is,

$$\hat{y}_s - t_{\alpha/2} \sqrt{s_{y|x}^2 \left\{ \frac{1}{n} + \frac{(x_s - \bar{x})^2}{\sum' xx} \right\}}, \qquad \hat{y}_s + t_{\alpha/2} \sqrt{s_{y|x}^2 \left\{ \frac{1}{n} + \frac{(x_s - \bar{x})^2}{\sum' xx} \right\}} \tag{54.4}$$

where $t_{\alpha/2} = t(n - 2); \alpha/2)$ and $\hat{y}_s = \bar{y} + b(x_s - \bar{x})$.

Notes: (i) Because of the $(x_s - \bar{x})^2$ term in (54.4), the interval gets wider the further x_s is from $\bar{x}$. In planning to predict a μ_s therefore, x-values chosen so that $\bar{x} = x_s$ will give the narrowest interval. (ii) The result (54.4) applies strictly for only one μ_s-value. If such intervals are required for several population means, $t_{\alpha/2}$ in (54.4) can be replaced by $\sqrt{2F_\alpha}$ where $F_\alpha = F(2, n - 2; \alpha)$; on this see Miller (1981).

The y-population mean at $x = 0$: As one special case of the above, the estimate of the population mean of y-values at the $x = 0$ origin is, from (54.1) with $x_s = 0$,

$$\bar{y} - b\bar{x} \tag{54.5}$$

which is the intercept of the estimation line (Fig. 48) and the estimate of the α-parameter in the alternative regression model $y_i = \alpha + \beta x_i + \varepsilon_i$. If it is required to test the significance, from zero, of the intercept, the test statistic, given by (54.3) with $\mu_s = 0$ in the numerator and $x_s = 0$ in the expression for $\hat{V}(\hat{y}_s)$, is referred to the $t(n-2)$-distribution. Alternatively, if H_A is two-sided, we can check to see if the value zero lies in the $100(1-\alpha)\%$ confidence interval for α which is, from (54.4) with $x_s = 0$,

$$(\bar{y} - b\bar{x}) - t_{\alpha/2}\sqrt{s_{y|x}^2\left(\frac{1}{n} + \frac{\bar{x}^2}{\sum' xx}\right)}, \qquad \bar{y} - b\bar{x} + t_{\alpha/2}\sqrt{s_{y|x}^2\left(\frac{1}{n} + \frac{\bar{x}^2}{\sum' xx}\right)} \tag{54.6}$$

Example: For the example in §52 ($n = 5$) it was found that $\bar{x} = 1.44$, $\bar{y} = 2.45$, $\sum' xx = 1.405$, $b = 1.97$, and $s_{y|x}^2 = 0.0038$, with three degrees of freedom.

The intercept estimate (54.5) is therefore

$$\bar{y} - b\bar{x} = 2.45 - 1.97(1.44) = -0.387$$

and the 99% confidence interval for its population mean is, from (54.6) with $t_{\alpha/2} = t(3; 0.01/2) = 5.841$,

$$-0.387 \mp 5.841\sqrt{0.0038\left(\frac{1}{5} + \frac{1.44^2}{1.405}\right)} = -0.85, 0.079$$

which includes zero as a statistically compatible value for the intercept.

Note: Estimation procedures for the model $y_i = \beta x_i + \varepsilon_i$ differ according to alternative assumptions available for $V(\varepsilon_i)$; statistical methods texts should be consulted for procedural details.

§55. PREDICTION INTERVALS FOR A "FUTURE," NOT-YET-OBSERVED, y-VALUE

It is sometimes required to obtain an interval, called a prediction interval, within which a "future" y-observation at some chosen x-value can be expected to fall—with specified confidence. The result, derived below, is that:

The $100(1-\alpha)\%$ **prediction interval** for a "future" y-observation, y_s say, at the specified x-value, $x = x_s$ say, is

$$\bar{y} + b(x_s - \bar{x}) \mp t(n-2; \alpha/2)\sqrt{s_{y|x}^2\left\{1 + \frac{1}{n} + \frac{(x_s - \bar{x})^2}{\sum' xx}\right\}} \tag{55.1}$$

Notes: (i) Comparison with (54.4) shows that the prediction interval will be wider than the corresponding interval for $\mu(y|x_s)$ because of the extra unity term multiplying $s^2_{y|x}$. (ii) If several, r say, y-observations are to be taken at the same *x-value* x_s, the 1 in Eq. (55.1) is replaced by $1/r$, to give a prediction interval for the mean of the r y-values. (iii) If prediction intervals are required *at k different x-values*, $x_{s1}, x_{s2}, \ldots, x_{sk}$, the critical value in (55.1) needs to be modified to $\sqrt{kF(k, n-2; \alpha)}$ (Miller, 1981).

Example: For the example in §52, find (i) a 99% confidence interval for the population mean $\mu(y|1.5)$ at $x = x_s = 1.5$; (ii) a 99% prediction interval for a future observation at $x_s = 1.5$; and (iii) a 99% prediction interval for the mean of $r = 2$ future observations at $x_s = 1.5$.

At $x_s = 1.5$, the estimation equation gives

$$\hat{y} = -0.39 + 1.97(1.5) = 2.565$$

Hence, from (54.4), the 99% confidence interval for $\mu(y|1.5)$ is

(i) $$2.565 \mp 5.841 \sqrt{0.0038\left\{\frac{1}{5} + \frac{(0.06)^2}{1.405}\right\}} = 2.40, 2.73$$

The 99% prediction interval for a future observation at $x_s = 1.5$ is, from (55.1),

(ii) $$2.565 \mp 5.841 \sqrt{0.0038\left\{1 + \frac{1}{5} + \frac{(0.06)^2}{1.405}\right\}} = 2.17, 2.96$$

The 99% prediction interval for the mean of two future observations at $x_s = 1.5$ is

(iii) $$2.565 \mp 5.841 \sqrt{0.0038\left\{\frac{1}{2} + \frac{1}{5} + \frac{(0.06)^2}{1.405}\right\}} = 2.26, 2.87$$

Derivation⁺ of (55.1): The predicted value $\hat{y}_s$ and the future observation y_s are independent, normally distributed, have the same population mean $\mu(y|x_s)$, and have the respective variances $\sigma^2_{y|x}\{1/n + (x_s - \bar{x})^2/\sum' xx\}$ and $\sigma^2_{y|x}$. Their difference $(\hat{y}_s - y_s)$ accordingly has population mean zero and its variance is estimated as

$$\hat{V}(\hat{y}_s - y_s) = \hat{V}(\hat{y}_s) + \hat{V}(y_s)$$

$$= s^2_{y|x}\left\{\frac{1}{n} + \frac{(x_s - \bar{x})^2}{\sum' xx}\right\} + s^2_{y|x}$$

$$= s^2_{y|x}\left\{1 + \frac{1}{n} + \frac{(x_s - \bar{x})^2}{\sum' xx}\right\} \qquad (55.2)$$

It follows that

$$\frac{(\hat{y}_s - y_s) - 0}{\sqrt{\hat{V}(\hat{y}_s - y_s)}} \quad \text{is a } t(n-2)\text{-variate}$$

so that

$$P\left[-t(n-2); \alpha/2) < \frac{\hat{y}_s - y_s}{\sqrt{\hat{V}(\hat{y}_s - y_s)}} < t(n-2); \alpha/2\right] = 1 - \alpha$$

Using (55.2) for the variance estimate and an argument like that in §13 then gives the result (55.1) above.

§56. CONFIDENCE INTERVALS FOR $\sigma_{y|x}^2$

The $100(1 - \alpha)\%$ confidence interval for $\sigma_{y|x}^2$ is from

$$\frac{\sum (y - \hat{y})^2}{\chi^2(n-2; \alpha/2)} \quad \text{to} \quad \frac{\sum (y - \hat{y})^2}{\chi^2(n-2; 1 - \alpha/2)} \tag{56.1}$$

where $\sum (y - \hat{y})^2$ is the ANOVA residual sum of squares.

The result follows by the argument in §15 because, with two constraints on the $(y_i - \hat{y}_i)$,

$$\frac{\sum (y_i - \hat{y}_i)^2}{\sigma_{y|x}^2} = \frac{(n-2)s_{y|x}^2}{\sigma_{y|x}^2} \quad \text{is a } \chi^2(n-2)\text{-variate}$$

Example: The residual sum of squares in Table 52.3 is 0.0114 with 3 df. With $\chi^2(3; 0.025) = 9.348$ and $\chi^2(3; 0.975) = 0.216$, the 95% confidence interval for $\sigma_{y|x}^2$ is from

$$\frac{0.0114}{9.348} \quad \text{to} \quad \frac{0.0114}{0.216} = 0.0012, 0.0528$$

Linear Correlation:
Measuring Relationship

§57. THE CORRELATION COEFFICIENT

Regression analysis is concerned with *prediction*, that is, one of two associated variables is selected for prediction from the other. In Model II regression situations (§47) it is also often required to assess the *strength of the relationship*, the *correlation*, between two continuously distributed variates. For example, taller students are generally heavier than smaller students; in this case height and weight are the two positively correlated variates because, subject to exceptions and variability, high values of one variate are associated with high values of the other and conversely. On the other hand, negative correlation generally obtains between the variates $x = $ city size and $y = $ air quality.

The correlation coefficient provides a numerical measure of the strength of the relationship or association between two variates. If a random sample of n experimental units is taken and two variate values, x and y, are measured on each unit, the data are the n pairs

$$(x_1, y_1), (x_2, y_2), \ldots, (x_n, y_n)$$

The sample correlation coefficient, r, is then defined as

$$r = \frac{1}{n-1} \sum_1^n \left(\frac{x_i - \bar{x}}{s_x}\right)\left(\frac{y_i - \bar{y}}{s_y}\right) = \frac{\sum (x_i - \bar{x})(y_i - \bar{y})}{(n-1)s_x s_y} \qquad (57.1)$$

Except that the sum is divided by $(n-1)$ rather than n, r is the average of the products $(x_i - \bar{x})(y_i - \bar{y})/s_x s_y$ of the x_i and y_i deviations from their respective sample means, each deviation being standardized into sample standard deviation units.

109

The sample covariance between the x and y variates is s_{xy} defined as

$$s_{xy} = \frac{\sum_1^n (x_i - \bar{x})(y_i - \bar{y})}{n-1} = \frac{\sum' xy}{n-1} \tag{57.2}$$

The sample covariance is analogous to the sample variance and is, in fact, the sample variance if $x = y$ in (57.2). From (57.1) and (57.2), the sample correlation coefficient can be written

$$r = \frac{s_{xy}}{s_x s_y} \tag{57.3}$$

The population correlation coefficient, ρ: If, conceptually, x and y values were to be measured on the whole infinite population of units, r in (57.1) would become the population correlation coefficient, denoted by ρ, where

$$\rho = \frac{\mu[(x - \mu_x)(y - \mu_y)]}{\sigma_x \sigma_y} \tag{57.4}$$

The correlation coefficient r is accordingly the sample estimate of ρ in (57.4). Correspondingly, the population mean of the sample covariance (57.2) is

$$\mu[s_{xy}] = \sigma_{xy} = \mu[r s_x s_y] = \rho \sigma_x \sigma_y \tag{57.5}$$

§58. SEVEN PROPERTIES OF CORRELATION COEFFICIENTS

(i) Because the correlations between x and y and between y and x are naturally the same, correlation coefficients (57.1) and (57.4) are symmetric in x and y; these can be interchanged without affecting r or ρ. In contrast, the denominator of a sample regression coefficient is $\sum' xx$ for the regression coefficient of y on x but is $\sum' yy$ for the regression coefficient of x on y.

(ii) The correlation coefficient is dimensionless, because, in (57.1), each deviation in the numerator has the same units as its standard deviation in the denominator.

(iii) The sign of the correlation coefficient is the same as the sign of the y-on-x or x-on-y regression coefficients; in all three the sign is that of the numerator sum $\sum_1^n (x_i - \bar{x})(y_i - \bar{y})$ because s_x and x_y are necessarily positive.

(iv) The maximum value of r and ρ is $+1$; their minimum value is -1. Maximum linear correlation obtains when the points (x_i, y_i), $i =$

$1, 2, \ldots, n$, all lie exactly on a straight line. If the slope of this line is b, then $(y_i - \bar{y}) = b(x_i - \bar{x})$, in which case

$$s_y^2 = b^2 s_x^2 \quad \text{and} \quad s_{xy} = \frac{\sum (x - \bar{x})b(x - \bar{x})}{n-1} = \frac{b \sum' xx}{n-1} = bs_x^2$$

Hence, from the definition (57.3),

$$r = \frac{bs_x^2}{\sqrt{s_x^2(b^2 s_x^2)}} = \pm 1$$

(v) The correlation coefficient is unaltered (except possibly for its sign) by linear conversion or coding operations. Thus, if we change from x and y to new variates u and v, where

$$u = cx + d \quad \text{and} \quad v = fy + g \tag{58.1}$$

and c, d, f, and g are any constant numbers, then for r_{uv}, the correlation coefficient between u and v,

$$\begin{aligned} r_{uv} &= r_{xy}, \quad \text{if } c \text{ and } f \text{ have the same sign} \\ r_{uv} &= - r_{xy}, \quad \text{if } c \text{ and } f \text{ have different signs} \end{aligned} \tag{58.2}$$

This result is obtained by noting that $(u_i - \bar{u}) = (a + cx_i) - (a + c\bar{x}) = c(x_i - \bar{x})$ so that, from (6.3), $s_u = |c|s_x$ and hence $(u_i - \bar{u})/s_u = c(x_i - \bar{x})/|c|s_x$. This, and the similar result for $(v_i - \bar{v})/s_v$, shows that

$$\frac{\sum (u - \bar{u})(v - \bar{v})}{s_u s_v} = \frac{cf \sum (x - \bar{x})(y - \bar{y})}{|c||f|s_x s_y}$$

so that, after dividing by $(n - 1)$, (58.2) is implied because

$$r_{uv} = \frac{cfr_{xy}}{|c||f|}$$

(vi) The correlation coefficient is not a good measure of anything but *linear correlation*. As an extreme demonstration suppose there are four (x, y) pairs $(-1, -1)$, $(-1, 1)$, $(1, -1)$, and $(1, 1)$. Because $\sum (x_i - \bar{x})(y_i - \bar{y}) = 0$, the correlation coefficient is then zero even though x and y are perfectly related by the equation $x_i^2 + y_i^2 = 2$, $i = 1, \ldots, 4$.

(vii) Correlation does not mean "causation;" a high correlation coefficient does not mean that x *causes* y or conversely. Thus, for the height and weight correlation, weight increases do not cause height increases or conversely. Instead, factors such as genetics and nutrition induce associated changes in both variates.

Figure 58 shows some characteristics of correlation coefficients.

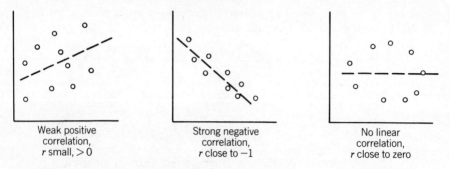

| Weak positive correlation, r small, > 0 | Strong negative correlation, r close to -1 | No linear correlation, r close to zero |

Figure 58. Correlation types.

§59. CALCULATING CORRELATION COEFFICIENTS

With $\sum' xy = \sum (x - \bar{x})(y - \bar{y})$ and $s_x^2 = \sum' xx/(n-1)$, Eq. (57.1) gives

$$r = \frac{\sum' xy/(n-1)}{\sqrt{s_x^2 s_y^2}} = \frac{\sum' xy/(n-1)}{\sqrt{\{\sum' xx/(n-1)\}\{\sum' yy/(n-1)\}}}$$

from which the $(n-1)$ cancels, leaving

$$r = \frac{\sum' xy}{\sqrt{(\sum' xx)(\sum' yy)}} \tag{59.1}$$

From (49.1)

$$\sum' xx = \sum x^2 - \frac{(\sum x)^2}{n} \quad \text{and} \quad \sum' xy = \sum xy - \frac{(\sum x)(\sum y)}{n}$$

so that (59.1) gives

$$r = \frac{\sum xy - (\sum x)(\sum y)/n}{\sqrt{\{\sum x^2 - (\sum x)^2/n\}\{\sum y^2 - (\sum y)^2/n\}}}$$

$$= \frac{n \sum xy - (\sum x)(\sum y)}{\sqrt{\{n \sum x^2 - (\sum x)^2\}\{n \sum y^2 - (\sum y)^2\}}} \tag{59.2}$$

Example: Find r for the data:

| x | 29 | 32 | 25 | 26 | 30 | $\sum x = 142$ |
| y | 36 | 45 | 29 | 28 | 34 | $\sum y = 172$ |

Computations:

$$\sum x^2 = 29^2 + \cdots + 30^2 = 4066, \qquad \sum y^2 = 36^2 + \cdots + 34^2 = 6102$$

$$\sum xy = 29(36) + \cdots + 30(34) = 4957$$

Hence, with $n = 5$ pairs, (59.2) gives

$$r = \frac{5(4957) - 142(172)}{\sqrt{\{5(4066) - (142)^2\}\{5(6102) - (172)^2\}}} = 0.92$$

Alternatively, in accordance with §58 (v), it can be checked that the same value for r can be computed from the pairs (u, v), $u = x - 30$, $v = y - 35$, which converts the (x, y) into the (u, v) pairs,

u	-1	2	-5	-4	0	$\sum u = -8$
v	1	10	-6	-7	-1	$\sum v = -4$

§60. BIVARIATE NORMAL DISTRIBUTIONS

A univariate probability density function (pdf) gives the probability that a randomly selected continuous variate, x or y say, will fall in some specified infinitesimal *interval* within the range of the variate. A bivariate pdf gives the probability that a point pair (x, y), randomly selected from the bivariate population, will fall in some specified infinitesimal *region* in the (x, y) plane. For a bivariate *normal* population, the pdf is an algebraic expression, in (x, y) and the population parameters, which shows that the probability is high for regions close to the mean of the distribution, the point (μ_x, μ_y), and becomes smaller the further the specified region is from (μ_x, μ_y). The probability density, $f(x, y)$ at the point (x, y), measured on an axis perpendicular to the (x, y) plane, is the height from the point to the bell-shaped surface shown in Fig. 60. The total volume under the surface is unity.

The values of five parameters, μ_x, μ_y, σ_x^2, σ_y^2, and ρ, are required to specify probability densities for a particular bivariate normal distribution. Estimates of these parameters, calculated from a random sample of (x, y) pairs, are $\bar{x}, \bar{y}, s_x^2$ from (2.8), s_y^2 similarly calculated from the y-values, and r from (59.2).

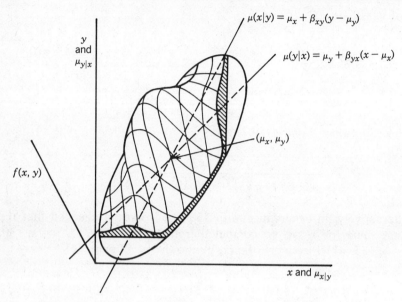

Figure 60. A bivariate normal distribution. Adapted from Hoel (1962).

§61. SOME PROPERTIES OF BIVARIATE NORMAL DISTRIBUTIONS

(i) At each individual x-value there is a normal distribution of y-values. The mean and variance of this distribution are $\mu_{y|x}$ and $\sigma_{y|x}^2 = \sigma_y^2(1 - \rho^2)$. As x changes, the $\mu_{y|x}$ all lie on the population y-on-x regression line, shown in Fig. 60:

$$\mu_{y|x} = \mu_y + \beta_{yx}(x - \mu_x) \qquad (61.1)$$

Taking $y_i = \mu(y|x_i) + \varepsilon_i$, and with the estimate $\bar{x}$ for μ_x, (61.1) provides the basis for the regression model (47.3) which gives b in (48.1) as the estimate of β_{yx} and $s_{y|x}^2$ in (51.3) as the estimate of $\sigma_y^2(1 - \rho^2)$.

(ii) At each individual y-value there is a normal distribution of x-values with mean $\mu_{x|y}$ and variance $\sigma_{x|y}^2 = \sigma_x^2(1 - \rho^2)$. As y changes, the $\mu_{x|y}$ all lie on the other population line in Fig. 60:

$$\mu_{x|y} = \mu_x + \beta_{xy}(y - \mu_y) \qquad (61.2)$$

The estimate of β_{xy} is the sample x-on-y regression coefficient $b_{xy} = \sum' xy / \sum' yy$ and the estimate of $\sigma_{x|y}^2$ is found from (51.3) by interchanging the x and y values.

(iii) It can be shown from the definitions of the quantities involved that

$$\beta_{yx} = \frac{\rho \sigma_y}{\sigma_x} \quad \text{and} \quad \beta_{xy} = \frac{\rho \sigma_x}{\sigma_y} \tag{61.3}$$

so that

$$(\beta_{yx})(\beta_{xy}) = \rho^2 \tag{61.4}$$

Correspondingly, in the sample case,

$$b_{yx} = \frac{rs_y}{s_x}, \qquad b_{xy} = \frac{rs_x}{s_y}, \qquad r^2 = (b_{yx})(b_{xy}) \tag{61.5}$$

For example,

$$\frac{rs_y}{s_x} = \frac{\sum' xy}{\sqrt{(\sum' xx)(\sum' yy)}} \frac{\sqrt{\sum' yy}}{\sqrt{\sum' xx}} = \frac{\sum' xy}{\sum' xx} = b_{yx}$$

It is seen that either of the two regression lines (61.1) and (61.2) can be estimated by the Model II regression procedures and the correlation coefficient is the geometric mean of the two regression coefficients. The choice between the lines in a practical regression situation is made by taking the variate to be predicted, or estimated, as y and the predictor variate as x.

(iv) Since β_{xy} is the tangent of the angle that the x-on-y line makes with the y-axis, the tangent of the angle this line makes with the x-axis is $1/\beta_{xy}$. We know that the x-on-y line is the steeper of the two for, otherwise, $1/\beta_{xy} < \beta_{yx}$ would give $(\beta_{xy})(\beta_{yx}) = \rho^2 > 1$! If the two lines have the same slope, we find $\rho^2 = 1$, the perfect correlation case. As the correlation decreases, the angle between the two lines becomes wider until, for a 90° angle, the correlation coefficient is zero, the y-on-x regression line is parallel to the x-axis, and the x-on-y line is parallel to the y-axis. Hence, if either the y-on-x or the x-on-y regression coefficient is zero so is the other and so is the correlation coefficient. The important result (v) then obtains.

(v) *Statistical independence:* (See §1.) If the population correlation coefficient between two bivariate normally distributed variates is zero, the two variates are statistically independent.

Note: Zero correlation coefficients do not always guarantee statistical independence but only when the two variates are *normally distributed*.

§62. THE CORRELATION COEFFICIENT AND MODEL II REGRESSION ANALYSIS

If $\hat{y}_i$ is the value predicted from (48.2) at $x = x_i$, then the square of the correlation coefficient calculated from the n pairs $(\hat{y}_i, y_i)$ is equal to the square of the correlation coefficient calculated from the original set of (x_i, y_i) pairs, that is,

$$r_{\hat{y}y}^2 = r_{xy}^2 \qquad (62.1)$$

Accordingly, since the closer the $\hat{y}_i$'s are to the observed y_i's, the better the regression line fits, or describes, the data, the success of the equation in this respect is nicely measured by r_{xy}^2. To obtain (62.1) we note that

$$\hat{y}_i = \bar{y} + b(x_i - \bar{x}) = bx_i + (\bar{y} - b\bar{x})$$

which is, with $c = b$, $d = (\bar{y} - b\bar{x})$, and $u = \hat{y}_i$, the first of the conversion equations in (58.1), the other being simply $v = 1(y) + 0$. Squaring the correlation coefficients in (58.2) establishes (62.1).

A useful property of the correlation coefficient is that its square plays the same role in Model II regression as that of the coefficient of determination (52.3) in Model I regression; that is,

$$\text{the sum of squares due to regression} = r^2 \left(\sum' yy \right) \qquad (62.2)$$

because the sum of squares due to regression, from Table 52.1, with (59.1), is

$$\frac{(\sum' xy)^2}{\sum' xx} = \frac{(\sum' xy)^2 (\sum' yy)}{(\sum' xx)(\sum' yy)} = r^2 \sum' yy$$

It follows that the residual deviations sum of squares $\sum (y - \hat{y})^2$ is $\sum' yy - r^2 \sum' yy = (1 - r^2) \sum' yy$ and hence that the regression ANOVA can equivalently be presented as in Table 62.

Table 62. ANOVA for a Linear Regression

Variability Source	df	ss	ms
Regression line	1	$r^2 \sum' yy$	$r^2 \sum' yy$
Residual deviations	$n - 2$	$(1 - r^2) \sum' yy$	$s_{y|x}^2 = (1 - r^2) \sum' yy/(n - 2)$
Total	$n - 1$	$\sum' yy$	

In confirmation of Note (v) §52, the mean squares in Table 62 show that strong x, y correlations, giving large r^2 values, achieve large reductions in the unexplained variance, from s_y^2 to $s_{y|x}^2 = s_y^2(1 - r^2)(n - 1)/(n - 2)$.

§63. TESTING (1): IS THE POPULATION CORRELATION COEFFICIENT ZERO OR NOT?

(i) **The ANOVA F-test:** Equation (61.3) shows that $\rho = 0$ implies that $\beta_{yx} = 0$, so that the test required here, of $H_T: \rho = 0$, $H_A: \rho \neq 0$; α, is exactly equivalent to the test of $H_T: \beta_{yx} = 0$, $H_A: \beta_{yx} \neq 0$; α, made in §53.

Example: The ANOVA for the data in the §59 example is given in Table 63, so that, because $F_T = 16.7$ exceeds $F(1, 3; 0.05) = 10.1$, we can immediately infer that $\rho \neq 0$ subject to an $\alpha = 0.05$ Type I error probability.

Table 63. Regression ANOVA for Data in §59

Source	df	ss	ms	F_T
Regression	1	157.0	157.0	16.7
Deviations	3	28.2	9.4	
Total	4	185.2		

(ii) **The F- and t-test statistics:** The above principles lead to the test statistic applicable when, instead of the ANOVA, only the sample correlation coefficient and the sample size are known. Division of the two mean squares in Table 62 shows that the above F-test statistic is, after canceling $\sum' yy$,

$$F_T = \frac{(n - 2)r^2}{1 - r^2} \tag{63.1}$$

again for comparison with $F(1, n - 2; \alpha)$.

Example: Test $H_T: \rho = 0$, $H_A: \rho \neq 0$; 0.05, given the data $r = 0.92$, $n = 5$ pairs.

Test statistic: In this case we calculate the test statistic from (63.1) to find, exactly as above, and for comparison with $F(1, 3; 0.05)$, that

$$F_T = \frac{3(0.92^2)}{1 - 0.92^2} = 16.7$$

One-sided H_A: When H_A is one-sided, the test statistic is

$$\sqrt{F_T} = t_T = \sqrt{\frac{(n-2)r^2}{1 - r^2}} \tag{63.2}$$

which, with appropriate sign, is referred to the $t(n-2)$-distribution.

Example: Test $H_1: \rho \geq 0$, $H_A: \rho < 0$; 0.05, given that $r = -0.6$ was calculated from $n = 12$ pairs.

We find from (63.2) that

$$t_T = \sqrt{\frac{10(0.36)}{0.64}} = 2.37$$

and H_A is indicated because $-2.37 < -t(10; 0.05) = -1.812$ or, equivalently, because $2.37 > +1.812$.

(iii) **Tabulated critical values:** These are sometimes used to give, by degrees of freedom $(n-2)$, the $\alpha = 0.05$ and $\alpha = 0.01$ values to be exceeded by $|r|$ for the acceptance of H_A. When few correlation coefficients are to be tested, the table use is only marginally more expeditious than procedure (ii) above from which exceedance probabilities can also be identified.

Notes: (i) The above and the following procedures rely on the assumption that r has been calculated from a random sample of bivariate normally distributed (x, y) pairs.

(ii) Because r is not a normally distributed variate, the above procedures do not lead to confidence interval estimations of ρ. For these see §65.

§64. TESTING (2): IS ρ EQUAL TO SOME SPECIFIED NONZERO VALUE OR NOT?

The §63 methods only apply when the test hypothesized value ρ_T is zero. When $\rho_T \neq 0$ we have the following alternatives.

(i) **When the number of pairs n is < 25:** Hypothesis tests are, strictly, best made using the special tables (David, 1930). For general practice, however, the next, approximate, procedure will often suffice even if n is somewhat less than 25.

(ii) **When n > 25:** R. A. Fisher showed that if an r-value is converted to a variate z' using the conversion equation

$$z' = \ln \sqrt{\frac{1+r}{1-r}} \tag{64.1}$$

where ln denotes the natural logarithm, then, as n increases, z' becomes, to a close approximation, distributed normally with the *population parameters*

$$\mu(z') = \zeta = \ln \sqrt{\frac{1+\rho}{1-\rho}} \tag{64.2}$$

$$V(z') = \sigma_{z'}^2 = \frac{1}{n-3} \tag{64.3}$$

Accordingly, to test

$$H_T: \rho = \rho_T \neq 0, \quad H_A: \rho \neq \rho_T; \quad \alpha$$

we can equivalently test

$$H_T: \zeta = \zeta_T, \quad H_A: \zeta \neq \zeta_T; \quad \alpha$$

where, from (64.2),

$$\zeta_T = \ln \sqrt{\frac{1+\rho_T}{1-\rho_T}} \tag{64.4}$$

Test statistic and critical region: When H_T is true, the test statistic

$$z_T = \frac{z' - \zeta_T}{\sqrt{V(z')}} = \sqrt{n-3}\,(z' - \zeta_T) \tag{64.5}$$

is a regular $z \sim N(0, 1)$ variate so that critical regions are obtained from the Table A2 for the standard normal distribution. As usual, if H_A is one-sided, the critical region is one-tailed. Special tables for the forward conversions (64.1) and (64.2), and for their inverses, are widely available. The tables are omitted here because, in the form equivalent to (64.5),

$$z_T = \sqrt{n-3}\,\ln \sqrt{\frac{(1+r)(1-\rho_T)}{(1-r)(1+\rho_T)}} \tag{64.6}$$

the z_T-calculation is easily sequenced on modern hand computers using the fact that $\sqrt{A}\,\ln B = \sqrt{(\ln B)^2 A}$.

Example: Suppose that $r = 0.78$ has been calculated from $n = 30$ bivariate normally distributed (x, y) pairs and it is required to test

$$H_T: \rho = 0.5, \quad H_A: \rho \neq 0.5; \quad 0.05$$

If conversion tables are available, we find

$$z' = 1.045 \quad \text{for } r = 0.78$$

$$\zeta_T = 0.549 \quad \text{for } \rho_T = 0.5$$

to give the test statistic (64.5) as

$$z_T = \sqrt{27}\,(1.045 - 0.549) = 2.58$$

Alternatively, from (64.6),

$$z_T = \sqrt{27} \ln \sqrt{\frac{(1.78)(0.5)}{(0.22)(1.5)}} = 2.58$$

The test statistic exceeds $z(0.025) = 1.96$ to indicate H_A and the exceedance probability is $2(0.005) = 0.01$.

§65. CONFIDENCE INTERVAL ESTIMATES FOR CORRELATION COEFFICIENTS

Corresponding to the previous test procedure, Fisher's result can be used to find approximate confidence intervals for the population correlation coefficient estimated by the sample value r. We first find z' corresponding to r and then the $100(1 - \alpha)\%$ confidence interval for $\mu(z') = \zeta$ as

$$z' \mp z_{\alpha/2}\sqrt{\frac{1}{n-3}} \tag{65.1}$$

using (64.3) and the standard normal distribution table. The two limits of the interval in (65.1), $\tilde{\zeta}_L$ and $\tilde{\zeta}_H$ say, are then converted back into the limits for the correlation coefficient ρ. Again special tables are available for the conversions, or alternatively z' can be obtained from (64.1) and the inverse conversion formula

$$\tilde{\rho} = \frac{e^{2\tilde{\zeta}} - 1}{e^{2\tilde{\zeta}} + 1} \tag{65.2}$$

is easily calculated to give the lower and upper limits for ρ corresponding to $\tilde{\zeta} = \tilde{\zeta}_L$ and $\tilde{\zeta} = \tilde{\zeta}_H$.

Example: Given $r = 0.78$ from $n = 30$ pairs (x_i, y_i), find a 95% confidence interval for ρ.

First, z' is obtained from the table or as

$$z' = \ln \sqrt{\frac{1 + 0.78}{1 - 0.78}} = 1.045$$

Second, the 95% limits for ζ are, from (65.1),

$$\tilde{\zeta}_L, \tilde{\zeta}_H = 1.045 \mp 1.96\sqrt{1/27} = 0.69, 1.42$$

Third, (65.2) gives

$$\tilde{\rho}_L = \frac{e^{2(0.69)} - 1}{e^{2(0.69)} + 1} = 0.60$$

and similarly $\tilde{\rho}_H = 0.89$. The 95% confidence interval for ρ is then from 0.60 to 0.89 and the interval excludes the value $\rho_T = 0.5$, consistently with the result of the test in §64.

§66. TESTING THE DIFFERENCE BETWEEN TWO CORRELATION COEFFICIENTS

Suppose that correlation coefficients r_1 and r_2 are calculated from independent random samples of n_1 and n_2 bivariate normally distributed pairs and that it is desired to test the hypothesis that the corresponding population correlation coefficients, ρ_1 and ρ_2 say, are the same. Corresponding to ρ_1 and ρ_2, (64.2) will define ζ_1 and ζ_2 and if $\rho_1 = \rho_2$ then $\zeta_1 = \zeta_2$. We accordingly convert each r to a z' and test H_T: $\zeta_1 = \zeta_2$, using a procedure closely analogous to that in §24. Formally, we proceed as follows:

Hypothesis specification: H_T: $\rho_1 = \rho_2$, H_A: $\rho_1 \neq \rho_2$; α. Equivalently H_T: $\zeta_1 = \zeta_2$, H_A: $\zeta_1 \neq \zeta_2$; α.

Data: Two independent random samples of n_1 and n_2 pairs give correlation coefficients of r_1 and r_2, respectively. We convert from r_1 to z'_1 and from r_2 to z'_2. That is, we find

$$z'_i = \ln \sqrt{\frac{1 + r_i}{1 - r_i}} \quad \text{for } i = 1, 2$$

Assumption: The procedure relies on the assumption that the two sets of pairs are independent and bivariate normally distributed.

Test statistic: The test statistic is $(z'_1 - z'_2)/\sqrt{V(z'_1 - z'_2)}$. To find $V(z'_1 - z'_2)$ we use the variance result (64.3) and the independence assumption to give

$$V(z'_1 - z'_2) = V(z'_1) + V(z'_2) = \frac{1}{n_1 - 3} + \frac{1}{n_2 - 3}$$

$$= \frac{n_1 + n_2 - 6}{(n_1 - 3)(n_2 - 3)}$$

so that the test statistic is

$$z_T = (z'_1 - z'_2) \sqrt{\frac{(n_1 - 3)(n_2 - 3)}{n_1 + n_2 - 6}} \tag{66.1}$$

which alternatively may be computed directly as

$$z_T = \sqrt{\frac{(n_1 - 3)(n_2 - 3)}{n_1 + n_2 - 6}} \ln \sqrt{\frac{(1 + r_1)(1 - r_2)}{(1 - r_1)(1 + r_2)}} \tag{66.2}$$

Critical region and inference: The two-sided H_A is accepted if $|z_T| > z_{\alpha/2}$ and rejected otherwise. The appropriate one-tailed critical region is used if H_A is one-sided.

Example: Suppose that a random sample of $n_1 = 20$ pairs from the first group gives $r_1 = 0.80$ and an independent random sample of $n_2 = 15$ pairs from a second group gives $r_2 = 0.62$.

Hypothesis specification: $H_T: \rho_1 = \rho_2$, $H_A: \rho_1 \neq \rho_2$;   0.05.

Data: Corresponding to $r_1 = 0.80$ and $r_2 = 0.62$ we find from (64.1)

$$z'_1 = \ln \sqrt{\frac{1.80}{0.20}} = 1.099, \qquad z'_2 = \ln \sqrt{\frac{1.62}{0.38}} = 0.725$$

Test statistic: From (66.1)

$$z_T = (1.099 - 0.725) \sqrt{\frac{17(12)}{29}} = 0.99$$

or, directly from r_1 and r_2 using (66.2),

$$z_T = \sqrt{\frac{17(12)}{29}} \ln \sqrt{\frac{1.80(0.38)}{0.20(1.62)}} = 0.99$$

Inference: Since $0.99 < z(0.025) = 1.96$, we have failed to detect any difference between the two population correlation coefficients.

§67. COMBINING TWO INDEPENDENT CORRELATION COEFFICIENTS

Again suppose that, from two independent random samples, correlation coefficients r_1, from n_1 pairs, and r_2, from n_2 pairs, have been calculated and that, on the belief that $\rho_1 = \rho_2 = \rho$, it is required to estimate the common correlation coefficient ρ. For this, we convert r_1 and r_2 to z_1' and z_2', combine these to estimate the common ζ, and then back-convert to the correlation scale.

Point estimation: With $z_i' = \ln \sqrt{(1 + r_i)/(1 - r_i)}$, $i = 1, 2$, a weighted average is taken to estimate ζ where the weights are the reciprocals of $V(z_1') = 1/(n_1 - 3)$ and $V(z_2') = 1/(n_2 - 3)$. The point estimate of ζ is then

$$\hat{\zeta} = \frac{(n_1 - 3)z_1' + (n_2 - 3)z_2'}{n_1 + n_2 - 6} \qquad (67.1)$$

which is simply $(z_1' + z_2')/2$ if $n_1 = n_2$. From (65.2) the corresponding point estimate of ρ is then

$$\hat{\rho} = \frac{e^{2\hat{\zeta}} - 1}{e^{2\hat{\zeta}} + 1} \qquad (67.2)$$

$100(1 - \alpha)$% Confidence interval for ρ: With the independence of z_1' and z_2', Rules 1 and 2 (§6, §7) can be applied to show that

$$V(\hat{\zeta}) = \frac{1}{n_1 + n_2 - 6}$$

so that the $100(1 - \alpha)$% confidence interval for ζ is

$$\hat{\zeta} \mp z_{\alpha/2} \sqrt{\frac{1}{n_1 + n_2 - 6}} = \hat{\zeta}_L, \hat{\zeta}_H \qquad (67.3)$$

These limits are reconverted using (65.2) to give $\hat{\rho}_L$ and $\hat{\rho}_H$.

Example: Estimate a common correlation coefficient from the data used in §66:

Group	n	r	z'
1	20	0.80	1.099
2	15	0.62	0.725

From (67.1)

$$\hat{\zeta} = \frac{17(1.099) + 12(0.725)}{29} = 0.944$$

From (67.3) the 95% confidence interval for ζ is

$$\hat{\zeta}_L, \hat{\zeta}_H = 0.944 \mp 1.96\sqrt{1/29} = 0.580, 1.308$$

Hence, using (67.2), the corresponding point estimate for ρ is

$$\hat{\rho} = \frac{e^{2(0.944)} - 1}{e^{2(0.944)} + 1} = 0.74$$

and similarly $\hat{\zeta}_L$ and $\hat{\zeta}_H$ convert to $(0.52, 0.86)$, the 95% confidence interval for ρ.

§68.[+] SUMS AND DIFFERENCES OF CORRELATED VARIATES

Suppose that x and y are two correlated variates with population means μ_x and μ_y, population variances σ_x^2 and σ_y^2, and population covariance $\rho\sigma_x\sigma_y$, as in (57.5) with ρ for the correlation coefficient estimated as r from n pairs $(x_1, y_1), \ldots, (x_n, y_n)$. Suppose also that c and d are two constants and that it is required to find the population mean and variance of the linear combination, $cx + dy$, of the two variates. The population mean is simply

$$\mu(cx + dy) = \mu(cx) + \mu(dy) = c\mu_x + d\mu_y \tag{68.1}$$

From its definition (5.1) the population variance is the mean of the squares of deviations such as

$$(cx + dy) - \mu(cx + dy) = cx + dy - c\mu_x - d\mu_y = c(x - \mu_x) + d(y - \mu_y)$$

Squaring this deviation gives

$$\{c(x - \mu_x) + d(y - \mu_y)\}^2 = c^2(x - \mu_x)^2 + d^2(y - \mu_y)^2 + 2cd(x - \mu_x)(y - \mu_y)$$

Hence, taking the population mean on both sides,

$$V(cx + dy) = c^2 \mu[(x - \mu_x)^2] + d^2 \mu[(y - \mu_y)^2] + 2cd\mu[(x - \mu_x)(y - \mu_y)]$$

Again, from the definition of population variance,

$$\mu[(x - \mu_x)^2] = \sigma_x^2 \quad \text{and} \quad \mu[(y - \mu_y)^2] = \sigma_y^2$$

and from the covariance definition (57.5)

$$\mu[(x - \mu_x)(y - \mu_y)] = \rho\sigma_x\sigma_y$$

Hence, the variance required is

$$V(cx + dy) = c^2\sigma_x^2 + d^2\sigma_y^2 + 2cd\rho\sigma_x\sigma_y \tag{68.2}$$

Notes: (i) The result (68.2) can be obtained by first squaring $(cx + dy)$ as

$$(cx + dy)^2 = c^2x^2 + d^2y^2 + 2cdxy$$

and then replacing x^2, y^2, and xy by σ_x^2, σ_y^2, and $\rho\sigma_x\sigma_y$, respectively.

(ii) If the two variates are uncorrelated then $\rho = 0$ and (68.2) reduces to

$$V(cx + dy) = c^2\sigma_x^2 + d^2\sigma_y^2 \tag{68.3}$$

which is just the result obtained by applying Rules 1 and 2 from §6 and §7 to the combination variate $(cx + dy)$: see §7, Example (i).

(iii) The mean (68.1) and variance (68.2) could be estimated, from a sample of n pairs, as the sample mean and variance of the n univariate quantities $(cx_1 + dy_1), \ldots, (cx_n + dy_n)$. The estimates are also respectively equal to $c\bar{x} + d\bar{y}$ and $c^2s_x^2 + d^2s_y^2 + 2rs_xs_y$, $(n-1)$ df.

Examples:

(i) Suppose that $\mu_x = 10$, $\mu_y = 15$, $\sigma_x^2 = 16$, $\sigma_y^2 = 25$, and $\rho = 0.8$. Find the population means and variances of the variates $u = 7x + 4y$ and $v = 7x - 4y$.

From (68.1) with $c = 7$ and $d = 4$,

$$\mu_u = 7\mu_x + 4\mu_y = 7(10) + 4(15) = 130$$

From (68.2)

$$\sigma_u^2 = V(u) = 7^2(16) + 4^2(25) + 2(7)(4)(0.8)(4)(5) = 2080$$

Similarly, with $c = 7$ and $d = -4$,

$$\mu_v = 7(10) - 4(15) = 10$$

$$\sigma_v^2 = 7^2(16) + (-4)^2(25) + 2(7)(-4)(0.8)(4)(5) = 288$$

It is noteworthy that, because of the positive correlation, σ_u^2 exceeds, and σ_v^2 is less than, the variance 1184, which (68.3) would give if x and y were uncorrelated.

(ii) If the correlation between two variates, y_A and y_B say, is ρ then for the variate $(y_A - y_B)$, taking $c = 1$ and $d = -1$ in (68.2),

$$V(y_A - y_B) = \sigma_A^2 + \sigma_B^2 - 2\rho\sigma_A\sigma_B \qquad (68.4)$$

Accordingly, if $\rho = 0$, $V(y_A - y_B) = \sigma_A^2 + \sigma_B^2$ while, if $\rho > 0$, $V(y_A - y_B) < \sigma_A^2 + \sigma_B^2$. The $\rho = 0$ case applies in the completely randomized experiment, §24; the $\rho > 0$ applies in the randomized pairs experiment, §23, in which good pairing will induce high positive correlation between the mates within pairs and hence will induce high precision by reducing the variance of the d-differences (and so of $\bar{d}$) in accordance with (68.4).

(iii) The expression in (54.2) for the variance of the estimate $\hat{y}_s$ of $\mu(y \mid x = x_s)$ in simple linear regression was obtained using the theoretical result that the estimates $\bar{y}$ and b, of the line parameters μ and β, are statistically independent. The procedures in this section can be used as follows to demonstrate the independence on the basis of the two results:

(a) that linear combinations of normally distributed variates are themselves normally distributed and
(b) that two normally distributed variates are statistically independent if their population correlation coefficient is zero, as noted in §61(v).

From the computational forms for the estimates, or the models (50.3) and (50.4) for them, it can be seen that requirement (a) is satisfied. With regard to (b) we note that (57.4) defines the numerator of the $(\bar{y}, b)$ correlation coefficient as $\mu[(\bar{y} - \mu)(b - \beta)]$, that is, as the population mean of the product $(\bar{y} - \mu)(b - \beta)$. With $\bar{\epsilon} = (\epsilon_1 + \epsilon_2 + \cdots + \epsilon_n)/n$, (50.3) and (50.4) show that this product is

$$\frac{1}{n}(\epsilon_1 + \epsilon_2 + \cdots + \epsilon_n) * \frac{1}{\sum' xx}\{(x_1 - \bar{x})\epsilon_1 + (x_2 - \bar{x})\epsilon_2 + \cdots + (x_n - \bar{x})\epsilon_n\}$$

which, if multiplied out, would consist of the sum of terms individually typified by

$$\frac{\epsilon_i}{n}\left[\frac{(x_j - \bar{x})\epsilon_j}{\sum' xx}\right] = \left[\frac{(x_j - \bar{x})}{n\sum' xx}\right]\epsilon_i\epsilon_j$$

for the various, equal and unequal, values of i and j from 1 to n. The numerator of $(\bar{y}, b)$ correlation coefficient is the sum of the population means of all these terms. Now, from the specification in the basic regression model (47.3) that the sample ϵ's are independent, it follows that the population mean of every individual term, such as $\{(x_2 - \bar{x})/n \sum' xx\}\epsilon_1\epsilon_2$, in which $i \neq j$, is zero. We are left therefore with the sum, $\sum_1^n [(x_i - \bar{x})/n \sum' xx]\epsilon_i^2$, of the n terms for which $i = j$. In each of these, the population mean of the ϵ_i^2 is just $\sigma_{y|x}^2$ by the definition of population variance. This leaves the sum $\sum_1^n \{(x_i - \bar{x})/n \sum' xx\}\sigma_{y|x}^2$ which is zero, because $\sum_1^n (x_i - \bar{x})$ is, and hence, their population correlation coefficient being zero, it follows from (b) that $\bar{y}$ and b are statistically independent.

Generalization: The result (68.2) is readily extended. For example, suppose there are three variates x_1, x_2, and x_3 with variances σ_1^2, σ_2^2, and σ_3^2, respectively, and with correlation coefficients ρ_{ij} between x_i and x_j. Suppose that constants c_1, c_2, and c_3 are taken to form the combination variate $u = c_1 x_1 + c_2 x_2 + c_3 x_3$. Then the rule in Note 1 above gives

$$V(u) = c_1^2\sigma_1^2 + c_2^2\sigma_2^2 + c_3^2\sigma_3^2 + 2c_1c_2\rho_{12}\sigma_1\sigma_2 + 2c_2c_3\rho_{23}\sigma_2\sigma_3 + 2c_1c_3\rho_{13}\sigma_1\sigma_3$$
(68.5)

§69. THE SPEARMAN RANK CORRELATION COEFFICIENT

When x and y are bivariate normally distributed, r in (59.2) provides the best estimate of the population correlation coefficient. An alternative estimate, which is not seriously inefficient with large sample sizes from the bivariate normal distribution and is recommended when the distribution cannot be specified, is the Spearman rank correlation coefficient. For this, data pairs (x_i, y_i) are replaced by the pairs (x_i', y_i'), $i = 1, 2, \ldots, n$, where x_i' is the rank of x_i in the x-sample and y_i' is the rank of y_i in the y-sample (ranked from 1, for the lowest, to n, for the highest). The rank correlation coefficient $r_{x'y'}$ can then be obtained by applying the usual definition (57.1) and formulas [(59.1) and (59.2)] to the (x_i', y_i') pairs. Equivalently, $r_{x'y'}$ can be conveniently calculated from the sum of the n squares of the rank differences, $d_i = x_i' - y_i'$, $i = 1, \ldots, n$, as

$$r_{x'y'} = 1 - \frac{6 \sum d_i^2}{n(n^2 - 1)}$$
(69.1)

Further details, including the treatment of ties and the special tables for hypothesis testing critical regions, are available in Conover (1980) and Hollander and Wolfe (1973).

Example: The data are the (x, y) pairs, from §59, in the first two columns of Table 69.

Table 69. Calculation of the Rank Correlation Coefficient

x	y	x'	y'	d	d^2
29	36	3	4	−1	1
32	45	5	5	0	0
25	29	1	2	−1	1
26	28	2	1	1	1
30	34	4	3	1	1
			Total	0	4

Since $x_3 = 25$ is the lowest in the x-series, it receives rank number $1 = x'_3$; the next highest x-value, $x_4 = 26$, is ranked second so that $x'_4 = 2$; and so on up to $x_5 = 32$ for which $x'_5 = 5$. The differences $d_i = x'_i - y'_i$ (column 5) sum to zero as they should and $\sum d_i^2 = 4$. Hence, from (69.1), the rank correlation coefficient is

$$r_{x',y'} = 1 - \frac{6(4)}{5(24)} = 0.8$$

Ranks are less informative than actual x, y values and the rank correlation coefficient does not achieve significance here whereas the $r = 0.92$ in §59 is significantly different from zero provided that the assumption that the data pairs are bivariate normally distributed is true.

Completely Randomized Experiments: Two or More Groups

§70. INVESTIGATION CONTEXTS

The methods in §24 are now extended to ANOVA procedures for investigating differences between the means of several populations from which samples of independent observations have been obtained. Typically, the sample data measure the effects of different groups, varieties, treatments, and so on, generically designated as groups for expository convenience. The abbreviation CR for completely randomized will also be used.

Randomization in controlled experiments: When the observations are to measure treatment effects on experimental units in controlled experiments, a randomization procedure (see Table A1) is employed to ensure that each unit has the same chance of receiving any one of the treatments. The aim of randomization is to avoid the possibilities of obtaining biased mean and variance estimates inherent in subjective or systematic procedures for assigning treatments to experimental units. In particular, randomization ensures that, if no treatment effects exist, the population variability between experimental units in different treatment groups is the same as that between the units within any treatment group. This is the basis for using the ratio of the between-treatments and within-treatment variances as the F-statistic (§75) for the detection of treatment differences.

Two situations are recognized, characterized by different objectives,

129

different statistical models, and different hypothesis specifications. Statistical Model I (§72) is appropriate when the group effects are unknown *fixed, parameter quantities*. Model II (§73) applies when the group effects are unknown *randomly distributed variates*.

§71. NOTATION FOR THE DATA

If n independent observations are obtained on the effects for each of a groups, the data can be arranged as in Table 71 wherein:

y_{ij} is the jth of the n replicate observations in the ith of the a groups,

$T_i = \sum_1^n y_{ij}$ is the total for the ith group

$\bar{y}_{i.} = \dfrac{T_i}{n} = \dfrac{\sum y_{ij}}{n}$ is the ith group mean

$G = \sum_1^a T_i = \sum_{i=1}^a \sum_{j=1}^n y_{ij}$ is the grand total of all the $N = na$ observations

$\bar{y}_{..} = \dfrac{G}{na} = \dfrac{\sum_i \sum_j y_{ij}}{na}$ is the overall mean

Table 71. Notation for the Data in a CR Experiment

Group	Replicate Observations 1	2	...	j	...	n	Total	Mean
1	y_{11}	y_{12}	...	y_{1j}	...	y_{1n}	T_1	$\bar{y}_{1.}$
2	y_{21}	y_{22}	...	y_{2j}	...	y_{2n}	T_2	$\bar{y}_{2.}$
$\vdots$			...				$\vdots$	$\vdots$
i	y_{i1}	y_{i2}	...	y_{ij}	...	y_{in}	T_i	$\bar{y}_{i.}$
$\vdots$			...				$\vdots$	$\vdots$
a	y_{a1}	y_{a2}	...	y_{aj}	...	y_{an}	T_a	$\bar{y}_{a.}$
				Grand total			G	$\bar{y}_{..}$ = overall mean

$\bar{y}_{..}$ is also the mean of the group means because

$$\bar{y}_{..} = \frac{1}{a}\sum_i \left(\frac{1}{n}\sum_j y_{ij}\right) = \frac{1}{a}\sum \bar{y}_i.$$

In the above notation a dot replaces a literal suffix which has been "summed (or overaged) out."

Example: From Table 75.2, where $a = 3$ and $n = 2$,

$$y_{11} = y_{A1} = 17.1, \qquad y_{12} = y_{A2} = 13.2, \qquad y_{32} = y_{C2} = 8.2$$

$$T_2 = T_B = 12.6 + 10.8 = 23.4, \qquad \bar{y}_{2.} = \bar{y}_B = \frac{23.4}{2} = 11.70$$

$$G = 30.3 + 23.4 + 17.7 = 71.4, \qquad \bar{y}_{..} = \frac{71.4}{2(3)} = 11.90$$

§72. MODEL I CR EXPERIMENTS: FIXED GROUP EFFECTS

Examples: (i) The airs above each of the $a = 8$ largest cities in one state are sampled on $n = 12$ independent occasions to compare the pollution concentrations of just these eight cities. (ii) For an experiment to compare $a = 4$ reading training methods, R_1, R_2, R_3, and R_4 say, for children, $n = 20$ schools are chosen at random to use method R_1, a further 20 are chosen at random to use R_2, and so on, so that each school in the population has the same initial chance of using any one of the four methods. Other allocation schemes, such as using the supposedly best method in the worst schools, could give biased estimates of any differences between the methods.

Statistical model: We first apply the simple statistical model in §9, with j instead of i now, and model the jth observation in the ith group as

$$y_{ij} = \mu_i + \varepsilon_{ij}, \qquad j = 1, \ldots, n; \quad i = 1, \ldots, a \qquad (72.1)$$

where μ_i is the unknown population mean parameter representing the ith group effect and ε_{ij} is the random variability component attending the observation y_{ij}.

Now with $\mu = \sum_1^a \mu_i/a$ as the overall mean of the group means and $\alpha_i = (\mu_i - \mu)$ as the deviation of μ_i from μ, we can substitute $\mu_i = \mu + (\mu_i - \mu) = \mu + \alpha_i$ into (72.1). Also, introducing the assumption for the distribution of ε_{ij} gives the complete model formulation as

$$y_{ij} = \mu + \alpha_i + \varepsilon_{ij}, \qquad \varepsilon_{ij} \sim NI(0, \sigma^2) \qquad i = 1, \ldots, a; \quad j = 1, \ldots, n$$
$$(72.2)$$

where it is to be noted that

$$\sum_i \alpha_i = \sum_i (\mu_i - \mu) = 0 \qquad (72.3)$$

Assumptions: The assumption $\varepsilon_{ij} \sim NI(0, \sigma^2)$ is required to validate hypothesis tests; further to the normality and independence, a particular feature is the requirement that the observations in all the groups have *the same population variance* $V(\varepsilon_{ij}) = \sigma^2$.

Hypothesis specification: As the objective we ask, "Are there or are there not, real differences between the groups effect parameters?" Accordingly, with α (not to be confused with the α_i) for the Type I error probability, the hypothesis specification is

$$H_T: \mu_1 = \mu_2 = \cdots = \mu_a = \mu, \quad H_A: H_T \text{ is false}; \qquad \alpha \quad (72.4)$$

or equivalently, based on the (72.2) model,

$$H_T: \text{Every } \alpha_i = 0, \quad H_A: \alpha_i \neq 0 \quad \text{for at least two } i\text{'s}; \qquad \alpha \quad (72.5)$$

The specifications (72.4) and (72.5) are commonly examined first. If H_T is true, the model reduces to $y_{ij} = \mu + \varepsilon_{ij}$ representing a single sample of na observations from just one population. If H_A is accepted and group differences are detected, it may then be asked—Which of the μ_i's are different? For this the LSD method (§78) or the procedures described in §99–§103 are available. Specifications for examining designated linear relationships between the μ_i's can also be constructed and tested as described in §94–§98.

Notes: (i) If the equal variance requirement is not met, heterogeneous residual variance can obscurely enhance the risks of false inferences from analyses. If, for $a > 2$, the truth of the assumption is doubtful, see, for example, Dowdy and Wearden (1983) for variance testing procedures; for $a = 2$ see §27; see §110 for some alleviating possibilities. (ii) If a Model I investigation were to be repeated with the same groups, the α_i's in (72.2), though unknown, would be unchanged. Contrast this with §73, Note (ii) for the random effects model.

§73. MODEL II CR EXPERIMENTS: RANDOM GROUP EFFECTS

Example: Acid sulfate concentrations are measured on 12 independent occasions at each of 20 locations *randomly selected* in a state so that $a = 20$ and $n = 12$. The objectives are (i) to estimate the average concentration for the state, (ii) to see if the variability between locations exceeds the within-location variability, and (iii) to obtain guidance on good choices for a and n for future investigations.

Statistical model: If y_{ij} is the jth of the n observations in the ith group, we write

$$y_{ij} = M_i + \varepsilon_{ij}, \qquad j = 1, \ldots, n; \quad i = 1, \ldots, a \qquad (73.1)$$

where M_i represents the unknown true value for the ith group and ε_{ij} is the random variability component of the y_{ij} observation.

As distinct from the Model I case, where each μ_i is a parameter, the M_i's are now random variates, with *population mean* μ say, the quantity it is desired to estimate. Writing the deviation of M_i from μ as $A_i = (M_i - \mu)$ gives an alternative formulation of (73.1) as

$$y_{ij} = \mu + A_i + \varepsilon_{ij}, \qquad j = 1, \ldots, n; \quad i = 1, \ldots, a \qquad (73.2)$$

Assumptions: The model is completed by three assumptions: (i) about the A_i, that

$$A_i \sim NI(0, \sigma_A^2) \qquad (73.3)$$

which replaces (72.3); (ii) about the ε_{ij}, that

$$\varepsilon_{ij} \sim NI(0, \sigma^2) \qquad (73.4)$$

and (iii) that the variates A_i and ε_{ij} are statistically independent.

The model represents each observation as attended by two sources of variability: one, the A_i, is constant within the group and changes from group to group; the other, ε_{ij}, changes from any one observation to any other.

Hypothesis specification: A primary objective of a Model II investigation is to obtain an economical, precise, estimate of the parameter μ in (73.2). To this end the choice of good values for a and n turns on the relative magnitudes of σ_A^2 and σ^2. In Model II therefore, the emphasis is on estimation of variances rather than hypothesis testing. It is of interest, however, to test

$$H_T: \sigma_A^2 = 0, \quad H_A: \sigma_A^2 > 0; \quad \alpha \qquad (73.5)$$

since, if $\sigma_A^2 = 0$, it suffices to take as many random observations as possible, ignoring the grouping.

Notes: (i) §72, Note (i) also applies in Model II situations. (ii) Since the A_i's in (73.2) are now random variables, they will represent different, unknown, numbers if repeated experiments are carried out in the same specific Model II situation.

§74. THE ANOVA IDENTITY FOR MODELS I AND II CR DATA

The deviation of any observation y_{ij} from the overall mean $\bar{y}_{..}$ can be split into the deviation of y_{ij} from its group mean $\bar{y}_{i.}$ and the deviation of the group mean from the overall mean because

$$(y_{ij} - \bar{y}_{..}) = (y_{ij} - \bar{y}_{i.}) + (\bar{y}_{i.} - \bar{y}_{..}) \qquad (74.1)$$

Correspondingly, it can be shown that $\sum_i \sum_j (y_{ij} - \bar{y}_{..})^2$, the sum of the squares of all the *na* deviations, can be expressed as

$$\sum_i \sum_j (y_{ij} - \bar{y}_{..})^2 = \sum_i \sum_j (y_{ij} - \bar{y}_{i.})^2 + n \sum_i (\bar{y}_{i.} - \bar{y}_{..})^2 \qquad (74.2)$$

which is known as the ANOVA identity. It shows that the total (corrected) sum of squares, $\sum_i \sum_j (y_{ij} - \bar{y}_{..})^2$, can be partitioned into the two added parts:

(i) $\sum_i \sum_j (y_{ij} - \bar{y}_{i.})^2$, which is the sum of the squares of the deviations of the observations *from their respective group means*, known as the within-groups sum of squares, and

(ii) $n \sum_i (\bar{y}_{i.} - \bar{y}_{..})^2$, which is n times the sum of the squares of the deviations of the group means *from the overall mean*. This quantity is known as the between-groups sum of squares.

Note[+]: The identity (74.2) is proved by squaring and adding both sides of (74.1) using

(i) $\sum_i \sum_j (\bar{y}_{ij} - \bar{y}_{i.})(\bar{y}_{i.} - \bar{y}_{..}) = \sum_i (\bar{y}_{i.} - \bar{y}_{..}) \sum_j (y_{ij} - \bar{y}_{i.}) = 0$

because $\sum_j (y_{ij} - \bar{y}_{i.}) = 0$ and

(ii) $$\sum_i \sum_j (\bar{y}_{i.} - \bar{y}_{..})^2 = n \sum_i (\bar{y}_{i.} - \bar{y}_{..})^2$$

§75. THE ARITHMETIC OF THE ANOVA AND A SIMPLE EXAMPLE

In the second part of (74.2), $\sum_{i=1}^{a} (\bar{y}_{i.} - \bar{y}_{..})^2$, is the sum of the squares of the a deviations $(\bar{y}_{i.} - \bar{y}_{..})$ which add to zero. Accordingly, we attribute $(a-1)$ degrees of freedom to the between-groups sum of squares.

Next, each of the $i = 1, \ldots, a$ quantities $\sum_{j=1}^{n} (y_{ij} - \bar{y}_{i.})^2$ is the sum of the squares of n deviations which add to zero. Altogether therefore, the within-group sum of squares will have $(n-1) + \cdots + (n-1) = a(n-1)$ degrees of freedom.

Similarly, the term on the left-hand side of (74.2) is the total sum of squares of an deviations, which add to zero, and therefore has $(an-1)$ degrees of freedom. And the degrees of freedom are additive, that is,

$$\begin{array}{ccc} \text{df(between-groups)} + \text{df(within-groups)} = \text{df(total)} \\ (a-1) \quad\quad + \quad\quad a(n-1) \quad\quad\quad an-1 \end{array} \tag{75.1}$$

Computationally convenient forms for the total and between-groups sums of squares are obtained using the identity $\sum_1^r (x_i - \bar{x})^2 = \sum x_i^2 - r\bar{x}^2$ from §2 with appropriate choices of x and r. Thus, noting that $\bar{y}_{..} = G/an$, the total sum of squares in (74.2) is, with $r = an$,

$$\sum \sum (y_{ij} - \bar{y}_{..})^2 = \sum \sum y_{ij}^2 - an\left(\frac{G}{an}\right)^2 = \sum \sum y_{ij}^2 - \frac{G^2}{an}$$

Similarly, noting that $\bar{y}_{i.} = T_i/n$, we have for the between-groups sum of squares, with $r = a$,

$$n \sum (\bar{y}_{i.} - \bar{y}_{..})^2 = n\left\{ \sum \left(\frac{T_i}{n}\right)^2 - a\left(\frac{G}{an}\right)^2 \right\} = \frac{\sum T_i^2}{n} - \frac{G^2}{an}$$

The within-group sum of squares is not calculated directly but obtained by the subtraction implied in (74.2).

Using (75.1) and the above computational forms, (74.2) can be arranged vertically to give the first three columns of the ANOVA, Table 75.1. The last column is completed by finding

$$s_B^2 = \frac{\text{ss between-groups}}{a-1}, \quad\quad s^2 = \frac{\text{ss within-groups}}{a(n-1)} \tag{75.2}$$

Table 75.1. ANOVA for a Completely Randomized Experiment
(a groups, n observations per group)

Variability Source	df	ss	ms
Between-groups	$a-1$	$\dfrac{\sum T_i^2}{n} - \dfrac{G^2}{an}$	$s_B^2 = \dfrac{ss}{a-1}$
Within-groups	$a(n-1)$	$\sum_i \sum_j (y_{ij} - \bar{y}_{i.})^2$	$s^2 = \dfrac{ss}{a(n-1)}$
Total	$an-1$	$\sum_i \sum_j y_{ij}^2 - \dfrac{G^2}{an}$	

Computation order: After columns 1 and 2 have been completed, the ANOVA calculations are: (i) obtain the correction term G^2/an; (ii) calculate the total ss; (iii) calculate the between-groups ss; (iv) obtain the within-groups ss by the subtraction

$$ss(\text{within-groups}) = ss(\text{total}) - ss(\text{between-groups})$$

and (v) calculate the mean squares by (75.2).

The continuation is the calculation of the variance ratio F-test statistic

$$F_T = \frac{\text{between-groups ms}}{\text{within-groups ms}} = \frac{s_B^2}{s^2} \tag{75.3}$$

and the appropriate H_A, from (72.5) or (73.5), is accepted if $F_T > F\{(a-1), a(n-1); \alpha\}$.

Example: The computations below Table 75.3 show how to get the ANOVA for the data in Table 75.2.

Table 75.2. Data from a CR Experiment ($a = 3$, $n = 2$)

Group	Observations		Total	Mean
A	17.1	13.2	30.3	15.15
B	12.6	10.8	23.4	11.70
C	9.5	8.2	17.7	8.85
			$G = 71.4$	$11.90 = \bar{y}_{..}$

Table 75.3. ANOVA for Table 75.2 Data

Source	df	ss	ms	F_T
Between-groups	2	39.81	19.905	5.93
Within-groups	3	10.07	3.357	
Total	5	49.88	9.98	

(i) Correction term $= \dfrac{G^2}{an} = 71.4^2/6 \qquad = 849.66$

(ii) $\sum\sum y_{ij}^2 = 17.1^2 + 13.2^2 + \cdots + 9.5^2 + 8.2^2 = 899.54$

$\qquad\qquad\qquad\qquad$ ss(total) $\qquad = 899.54 - 849.66 = 49.88$

(iii) $\sum \dfrac{T_i^2}{n} = (30.3^2 + 23.4^2 + 17.7^2)/2 \qquad = 889.47$

$\qquad\qquad\qquad$ ss(between-groups) $= 889.47 - 849.66 = 39.81$

(iv) $\qquad\qquad\qquad$ ss(within-groups) $\quad = 49.88 - 39.81 = 10.07$

(v) $\qquad\qquad\qquad$ ms(between-groups) $= 39.81/2 \qquad = 19.905$

$\qquad\qquad\qquad\qquad$ ms(within-groups) $\ = 10.07/3 \qquad = 3.357$

The statistic $F_T = 19.905/3.357 = 5.93$ which is less than $F(2, 3; 0.05) = 9.55$ so that no evidence of intergroup differences is obtained at $\alpha = 0.05$.

Note: Computer routines are available for obtaining the analysis. In one multiple regression approach the uncorrected sum of squares $\sum_i \sum_j y_{ij}^2$ is analyzed into a one degree of freedom ss for the overall mean and appropriate sums of squares, corresponding to those above, for the between-groups and within-group components. The Table 75.1 values are readily obtained from such computer outputs.

§76. THE BASIS OF THE ANOVA *F*-TEST

(i) On the assumption that $V(\varepsilon_{ij}) = \sigma^2$, a independent estimates of σ^2, one from each group, can be obtained, each with $(n-1)$ degrees of freedom, as

$$s_1^2 = \frac{\sum_{j=1}^n (y_{1j} - \bar{y}_1.)^2}{n-1}, \quad \ldots, \quad s_a^2 = \frac{\sum_{j=1}^n (y_{aj} - \bar{y}_a.)^2}{n-1}$$

Extending the $a = 2$ pooled variance estimate procedure in §24, the best estimate s^2 is again obtained by pooling the separate estimates to give

$$s^2 = \frac{\sum_{j=1}^{n} (y_{1j} - \bar{y}_{1.})^2 + \cdots + \sum_{j=1}^{n} (y_{aj} - \bar{y}_{a.})^2}{(n-1) + \cdots + (n-1)} = \frac{\sum_{i=1}^{a} \sum_{j=1}^{n} (y_{ij} - \bar{y}_{i.})^2}{a(n-1)}$$

(76.1)

with $a(n-1)$ degrees of freedom. The numerator of s^2 in (76.1) is the within-group sum of squares, the first term on the right-hand side of (74.2), so that s^2 itself is obtained in the ANOVA as the within-groups mean square.

(ii) Suppose that, from one population of y-observations, with $y_{ij} = \mu + \varepsilon_{ij}$ and population variance $V(\varepsilon_{ij}) = \sigma^2$, repeated random samples, each of n observations, are taken and the sample means $\bar{y}_{1.}, \bar{y}_{2.}, \ldots$ are calculated. These sample means will themselves be members of a population of which the population variance is σ^2/n, (8.2). If just a such sample means are obtained, and their mean is

$$\bar{y}_{..} = \frac{\sum_i \bar{y}_{i.}}{a} = \frac{\sum_i \sum_j y_{ij}}{an}$$

the quantity $[1/(a-1)] \sum_{i=1}^{a} (\bar{y}_{i.} - \bar{y}_{..})^2$ provides an estimate of σ^2/n with $(a-1)$ degrees of freedom. Multiplication by n shows that an estimate of σ^2 itself can be calculated as s_B^2 (B for "between") where

$$s_B^2 = \frac{n \sum_i (\bar{y}_{i.} - \bar{y}_{..})^2}{a-1}$$

(76.2)

The numerator of s_B^2 in (76.2) is the between-groups sum of squares, the second term on the right-hand side of (74.2), and s_B^2 is the ANOVA between-groups mean square.

We now regard the above a sample means as the means of the groups in a Model I or II completely randomized investigation. Then if, instead of $y_{ij} = \mu + \varepsilon_{ij}$ representing the H_T situation, H_A is true and real differences between the groups exist, the $\bar{y}_{i.}$'s will tend to be more scattered than sample means from just one population and s_B^2 in (76.2) will then estimate some quantity exceeding σ^2. Accordingly, the variance ratio

$$F_T = \frac{s_B^2}{s^2}$$

(76.3)

can be used as a test statistic for the hypothesis specifications (72.5) and (73.5) since, if H_T is true, F_T will probably be close to unity while, if H_A is true, F_T will probably exceed unity. Furthermore, s_B^2 and s^2 have $(a-1)$ and $a(n-1)$ degrees of freedom, respectively, and the outcome is that: (i) when the appropriate H_T in (72.5) or (73.5) is true, and (ii) when the ε_{ij}'s are normally distributed, the variance ratio F_T in (76.3) is an F-variate with $(a-1)$ numerator and $a(n-1)$ denominator degrees of freedom. Critical values can therefore be obtained as F_α from the one-tail tabulation such that for Type I error probability α,

$$P[F\{a-1, a(n-1)\} > F_\alpha] = \alpha \tag{76.4}$$

The single-tail F_α is appropriate because, from the above argument, H_T is to be accepted unless s_B^2 (too surprisingly) *exceeds* s^2.

§77. POPULATION MEANS OF THE ANOVA MEAN SQUARES

If repeated completely randomized experiments were to be carried out in the same investigational context, the data, and hence the ANOVA mean squares, would differ from one experiment to another. Conceptually, therefore, there are two populations, one of between-groups mean squares and one of within-group mean squares. The mean squares calculated in a particular experiment are estimates of the population means of these two populations. For these, the important results can be obtained that:

For both Model I and Model II, the population mean of the within-group mean square is just

$$\mu(s^2) = V(\varepsilon_{ij}) = \sigma^2 \tag{77.1}$$

For Model I, $y_{ij} = \mu + \alpha_i + \varepsilon_{ij}$, the population mean of the between-groups mean square is

$$\mu(s_B^2) = \sigma^2 + \frac{n \sum_i \alpha_i^2}{a-1} \tag{77.2}$$

For Model II, $y_{ij} = \mu + A_i + \varepsilon_{ij}$, where $V(A_i) = \sigma_A^2$, the population mean of the between-groups mean square is

$$\mu(s_B^2) = \sigma^2 + n\sigma_A^2 \tag{77.3}$$

The above results are collected in Table 77. They quantitatively confirm the argument developed in §76. For if H_T is true in Model I, the α_i's are all zero and hence $\sum \alpha_i^2$ in (77.2) is zero so that the within-groups and between-groups mean squares both estimate just σ^2 and the test statistic is expectedly close to unity. Conversely, when H_A is true, the α_i's are not all zero and $\sum \alpha_i^2 \neq 0$; the between-groups mean square will then estimate something exceeding σ^2 by the amount $n \sum \alpha_i^2 /$

Table 77. Population Means of ANOVA Mean Squares

Variability Source	df	ms	μ(ms) Model I	μ(ms) Model II
Between-groups	$a-1$	s_B^2	$\sigma^2 + \dfrac{n \sum \alpha_i^2}{a-1}$	$\sigma^2 + n\sigma_A^2$
Within-groups	$a(n-1)$	s^2	σ^2	σ^2

$(a-1)$ and the test statistic will expectedly exceed unity. It may be noted also that, since it appears in the numerator, the larger n is the better for the detection of intergroup effects. The same argument applies in the Model II case, that is, if $\sigma_A^2 > 0$ the between-groups mean square in the numerator of F_T estimates something exceeding σ^2 by $n\sigma_A^2$ so that F_T will expectedly exceed unity unless $H_T: \sigma_A^2 = 0$ is true.

§78. COMPLETELY RANDOMIZED EXPERIMENTS MODEL I: FURTHER ANALYSIS

(i) $F_T < F_\alpha$: If the outcome from the ANOVA of a completely randomized experiment is that the test statistic F_T in (76.3) is less than the critical value $F_\alpha = F\{a-1, a(n-1); \alpha\}$, the immediate inference favors the test hypothesis that the α_i were all zero or equivalently that there were no differences between the group means, the μ_i's in (72.1). A preferred inference is that, perhaps due to inadequate replication in the presence of large variability, the experiment was not sufficiently sensitive to detect group differences. Wide confidence intervals for population means and differences between two population means will support the latter inference.

The **100(1 − α)% confidence interval for the population mean μ_i is**

$$\bar{y}_{i.} \mp t\{a(n-1); \alpha/2\} \sqrt{\frac{s^2}{n}} \tag{78.1}$$

The **100(1 − α)% confidence interval for the difference between two population means ($\mu_i - \mu_k$) is**

$$(\bar{y}_{i.} - \bar{y}_{k.}) \mp t\{a(n-1); \alpha/2\} \sqrt{\frac{2s^2}{n}} \tag{78.2}$$

The results (78.1) and (78.2) are analogous to and based on the same arguments as those in §13, for $a = 1$, and §24, for $a = 2$; see also the general statements (13.5) and (13.6). The points to note here are that the variance estimate used is the ANOVA mean square within groups, s^2, and not the pooled estimate from just s_i^2 and s_k^2, and, correspondingly, that the multiplying t-value is taken from the t-distribution with $a(n-1)$ degrees of freedom. The support for this is the equal population variance assumption which allows the variability estimate from all the observations to be used to sharpen particular confidence intervals.

Example: From the ANOVA in Table 75.3, $s^2 = 3.357$ with $a(n-1) = 3$ df, so that $t(3; 0.05/2) = 3.182$. Hence, (78.1) gives the 95% confidence interval for $\mu_1 = \mu_A$ as

$$15.15 \mp 3.182 \sqrt{\frac{3.357}{2}} = 11.0, 19.3$$

and the 95% confidence interval for $(\mu_1 - \mu_2) = (\mu_A - \mu_B)$ is, from (78.2),

$$(15.15 - 11.70) \mp 3.182 \sqrt{\frac{2(3.357)}{2}} = 3.45 \mp 5.83 = -2.38, 9.28$$

Confidence intervals for other differences are similarly obtained; thus, the 95% confidence interval for $(\mu_B - \mu_C)$ is from -2.98 to 8.68.

(ii) $F_T > F_\alpha = F\{a-1, a(n-1); \alpha\}$: *Confidence intervals* for a μ_i and a $(\mu_i - \mu_k)$ difference are again obtained from (78.1) and (78.2). Furthermore, with the inference that not all the group population means are the same, the question arises—Which means differ? Several diagnostic procedures to examine this question are presented in §99–103. Of these, the most convenient and the one most sensitive to the detection of real intergroup effects is the following.

The protected least significant difference (LSD) method: For this, the α-level least significant difference is defined as

$$\text{LSD} = t\{a(n-1); \alpha/2\} \sqrt{\frac{2s^2}{n}} \qquad (78.3)$$

and two population means, μ_i and μ_k say, are inferred to be different only if the absolute value of the difference between their estimates exceeds the LSD, that is, if

$$|\bar{y}_{i.} - \bar{y}_{k.}| > t\{a(n-1); \alpha/2\} \sqrt{\frac{2s^2}{n}} \qquad (78.4)$$

When a is large it is convenient to rank the estimates in ascending order and test the successive differences to see whether or not they exceed the LSD yardstick as exemplified in §79.

The rationale for the procedure is the simple t-test that H_A: $\mu_i - \mu_k \neq 0$ is to be accepted if $|\bar{y}_{i.} - \bar{y}_{k.}|/\sqrt{2s^2/n} > t\{a(n-1); \alpha/2\}$ for a prescribed Type I error rate α. Correspondingly, the confidence intervals (78.2) will exclude zero if the LSD test indicates that the two means differ. In fact, the α-level LSD is equal to the semiwidth of the $100(1 - \alpha)\%$ confidence interval for the difference between two means.

Notes: (i) The, then unprotected, LSD test is sometimes used when $F_T < F_\alpha$. The possibility of making Type I errors however, is increased, especially if a is large, because altogether there are $\binom{a}{2} = a!/2!(a-2)!$ pairs which can be tested with corresponding "opportunities" for differences exceeding the LSD to occur by chance when H_T is true.

(ii) When $F_T < F_\alpha$ the larger differences between the more extreme means are sometimes deliberately selected for testing as possible indications of perhaps unanticipated population differences. It is inappropriate to use the LSD in such cases because its rationale is based on the test hypothesis that all the mean differences are zero. The population mean of *deliberately selected* large differences will exceed zero and correspondingly the critical distance for significance should exceed the LSD. Further experimentation can be recommended to re-examine the groups giving interestingly large selected differences. See also §103.

(iii) Individual t-tests can be made to examine differences between means and combinations of means specifically planned, in advance of data inspection, to measure interesting between-groups effects; see §95ff. Such tests have validity whether or not $F_T > F_c$.

§79. EXAMPLE ANALYSIS OF A MODEL I, COMPLETELY RANDOMIZED, EXPERIMENT

Ramsay (1953) reported the data in Table 79.1 from an experiment of which the objective, indicating a Model I situation, was to examine the possibility that the sodium concentration in the hemolymph of mosquito larvae could adjust to different external salt concentrations. Hemolymph concentrations were measured in larvae exposed to one of the $a = 3$ external salt concentrations: A = distilled water, B = 85 mM/liter NaCl, C = 1.7 mM/liter NaCl

Table 79.1. Effects of Different External Media on the Composition of Mosquito Larval Hemolymph

Treatment	Replicate Observations (Na in mol. equiv./liter)						Total	Mean
A	89	89	75	91	85	93	522	87.0
B	120	115	112	100	115	119	681	113.5
C	96	90	107	103	100	106	602	100.3
					Grand total		1805	100.28

A preliminary inspection of the data suggests that the normality and independence assumptions are reasonable. Since the number of observations was the same in each group, the sample ranges can be used to assess the equal variance assumption. The ranges, $93 - 75 = 18$, 20 and 17, for A, B and C, respectively, are satisfactorily close (had they not been, the separate estimates s_A^2, s_B^2 and s_C^2, could be calculated and tested for variance heterogeneity). The procedures in §75 for $a = 3$, $n = 6$ then give the ANOVA in Table 79.2.

Table 79.2. ANOVA for Table 79.1 Data

Variability Source	df	ss	ms	F_T
Between treatments	2	2106.78	1053.39	23.3[a]
Within treatments	15	678.83	45.25	
Total	17	2785.61		

[a] $P < 0.001$.

The test statistic for H_T: $\mu_A = \mu_B = \mu_C$, H_A: H_T is false is $F_T = 1053.39/45.25 = 23.3$ as shown in Table 79.2 where, for the rejection of H_T, it is indicated that the exceedance probability is less than 0.001 because $F_T > F(2, 15; 0.001) = 11.339$. The rejection of H_T suggests that some group mean differences exist and that the protected LSD diagnosis is applicable. For $\alpha = 0.01$, the LSD, from (78.3) with $t(15; 0.005) = 2.947$, is

$$\text{LSD} = 2.947 \sqrt{\frac{2(45.25)}{6}} = 11.45$$

The ordered means are $\bar{y}_A = 87$, $\bar{y}_C = 100.3$, and $\bar{y}_B = 113.5$ and, because both $(\bar{y}_C - \bar{y}_A) = 13.3$ and $(\bar{y}_B - \bar{y}_C) = 13.2$ exceed 11.45, we can infer real differences between μ_A and μ_C, between μ_B and μ_C, and, of course, between μ_A and μ_B.

As examples, the 99% confidence interval for μ_A is, from (78.1),

$$87.0 \mp 2.947 \sqrt{\frac{45.25}{6}} = 78.9, 95.1 \text{ Na units}$$

and the 99% confidence interval for $(\mu_B - \mu_A)$ is, from (78.2),

$$(113.5 - 87.0) \mp 2.947 \sqrt{\frac{2(45.25)}{6}} = 26.5 \mp \text{LSD}(\alpha = 0.01)$$

$$= 26.5 \mp 11.45 = 15.0, 38.0 \text{ Na units}$$

The estimated standard deviation of a mean and the coefficient of variation could also be calculated as $\sqrt{45.25/6} = 2.75$ and $100\sqrt{45.25}/100.28 = 6.7\%$ to give Table 79.3 with Table 79.2 for presentation.

Table 79.3. Mean Larval Hemolymph Sodium Concentrations (Na units)

Treatment	A	B	C	sd (mean)	Coefficient of Variation
Mean	87.0	113.5	100.3	2.75	6.7%

Note: Although it is convenient to apply the LSD to the ordered means, as above, the appropriate H_A's in LSD diagnoses are nevertheless two-sided as (78.4) suggests. A noteworthy outcome from the example is the suggestion that larval salt concentration increases with external salt concentration. Model I regression investigations could, with initially less unevenly spaced external concentrations as the x-values, be designed to investigate this in further experiments. A significantly positive linear regression coefficient, for example, would then imply that $\mu(y|x_2) > \mu(y|x_1)$ for $x_2 > x_1$.

§80. COMPLETELY RANDOMIZED EXPERIMENTS MODEL II: FURTHER ANALYSIS

(i) $F_T < F_\alpha = F\{(a-1), a(n-1); \alpha\}$: Here the direct inference is that $\sigma_A^2 = 0$. If, presuming no Type II error, we take $\sigma_A^2 = 0$ in fact, the A_i in the model (73.2) are then all zero, and, in the absence of intergroup variability, the observations can be regarded as a single sample of $N = an$

observations. Accordingly, the original model now simply reduces to

$$Y_{ij} = \mu + \delta_{ij} \tag{80.1}$$

where δ_{ij} replaces $(A_i + \varepsilon_{ij})$ with the assumption $\delta_{ij} \sim NI(0, \sigma^2)$. Recalling that an objective of a Model II investigation is the estimation of the overall mean μ, the point estimate is $\bar{y}_{..}$ and the $100(1-\alpha)\%$ confidence interval estimate is just, from (13.5),

$$\bar{y}_{..} \mp t(an-1; \alpha/2) \sqrt{\frac{s_y^2}{an}} \tag{80.2}$$

where $s_y^2 = \sum_i \sum_j (y_{ij} - \bar{y}_{..})^2/(an-1)$ is the mean square obtained from the total row of the ANOVA.

Example: Supposing that the data in Table 75.2 were obtained in a Model II investigation, $s_y^2 = 49.88/5 = 9.98$, with five degrees of freedom, is obtained from Table 75.3, so that the point and 95% confidence interval estimates for μ and $\bar{y}_{..} = 11.90$ and, from (80.2) with $t(5; 0.025) = 2.571$, $11.90 \mp 2.571 \sqrt{9.98/6}$.

(ii) $F_T > F_\alpha$: When $F_T > F_\alpha$ and the inference that $\sigma_A^2 > 0$ is accepted, $V(\bar{y}_{..})$ depends on both σ_A^2 and σ^2. In this case, the point estimate of μ is again $\bar{y}_{..}$:
   The $100(1-\alpha)\%$ confidence interval for μ is

$$\bar{y}_{..} \mp t(a-1; \alpha/2) \sqrt{\frac{s_B^2}{na}} \tag{80.3}$$

where s_B^2 is the between-groups mean square in the ANOVA.

Derivation[+]: Using the model expression for y_{ij} in (73.2) we find the model for $\bar{y}_{..}$ as

$$\bar{y}_{..} = \frac{\sum_i \sum_j y_{ij}}{an} = \frac{\sum_i \sum_j (\mu + A_i + \varepsilon_{ij})}{an}$$

$$= \frac{an\mu + n \sum_i A_i + \sum_i \sum_j \varepsilon_{ij}}{an} = \mu + \bar{A}_. + \bar{\varepsilon}_{..} \tag{80.4}$$

so that, with the independence of the A_i and ε_{ij},

$$V(\bar{y}_{..}) = V(\bar{A}_.) + V(\bar{\varepsilon}_{..}) = \frac{\sigma_A^2}{a} + \frac{\sigma^2}{an} = \frac{\sigma^2 + n\sigma_A^2}{an} \tag{80.5}$$

The numerator in (80.5) is the population mean of the ANOVA between-groups mean square s_B^2, see (77.3), so that the estimate of $V(\bar{y}_{..})$ is just s_B^2/an, with $(a-1)$ degrees of freedom. The result (80.3) then follows from (13.5).

§81. EXAMPLE ANALYSIS OF A COMPLETELY RANDOMIZED MODEL II INVESTIGATION

In 1973 David Voigt made repeated counts of midges in several zones of an Iowa marsh. It was found that the variability between the replicate counts increased as the mean zone count increased, negating the assumption of constant within-zone variability. A standard approach in such situations is to find a new variate that is directly related to the one observed and does satisfy the constant variance assumption; see §110. Here the square root conversion or transformation was selected and, before analysis, each count was replaced by its correspondent $y = \sqrt{\text{count} + 3/8}$.

For our present purposes, only part of the investigation is used. The $n = 4$ y-values for each of $a = 3$ zones are given in Table 81.1; the ANOVA is given in Table 81.2.

Table 81.1. Transformed Midge Count Data

Zone	Replicate y-Values				Total
	1	2	3	4	
1	6.43	3.22	5.86	4.40	19.91
2	11.89	10.36	7.24	12.54	42.03
3	6.95	4.05	10.02	8.97	29.99
				Grand total	91.93

Table 81.2. ANOVA for Table 81.1 Data

Variability Source	df	ss	ms	F_T
Between-zones	2	61.3219	30.6610	6.30[a]
Within-zones	9	43.7678	4.8631	
Total	11	105.0897		

[a] $0.01 < P < 0.025$.

The exceedance probability lies between 0.01 and 0.025 to indicate H_A because $F(2, 9; 0.025) = 5.715$ and $F(2, 9; 0.01) = 8.02$ bracket the test statistic value $F_T = 6.30$.

If the zones were deliberately selected special areas, according to vegetation type, for example, this would be a Model I situation and the interpretation would be that appreciable differences existed between counts for the particular zones studied. Here it is assumed that the zones were randomly selected from the whole marshland area so that Model II applies and hence the ANOVA test indicates that σ_A^2 is not zero. Then,

The point estimate of μ is $91.93/12 = 7.661$.

The 95% interval estimate (80.3) is, with $t(2; 0.05/2) = 4.303$,

$$7.661 \mp 4.303 \sqrt{\frac{30.6610}{12}} = 0.783, 14.539$$

The estimates can be converted back into count units by "undoing" the transformation. If $y = \sqrt{x + 0.375}$, x being the actual count, then $x = y^2 - 0.375$ and, accordingly, the point estimate of the population mean count is $7.661^2 - 0.375 = 58.3$ and the 95% confidence interval for the population mean count is

$$(0.783)^2 - 0.375, (14.539)^2 - 0.375 \approx 0, 211$$

§82. PLANNING FOR PRECISION: RELATIVE EFFICIENCY

If $\sigma_A^2 > 0$, the ANOVA results from one Model II experiment can be used to plan a second experiment to give a more precise estimate of the overall mean of interest, μ, by changing the number of groups from a to a_2 say, and the number of observations per group from n to n_2. The variance of the mean estimate $\bar{y}_{..2}$ in the second experiment would then be estimated from (80.5) as

$$\hat{V}(\bar{y}_{..2}) = \frac{\hat{\sigma}_A^2}{a_2} + \frac{\hat{\sigma}^2}{a_2 n_2} \tag{82.1}$$

where $\hat{\sigma}_A^2$ and $\hat{\sigma}^2$ are estimates of the variance components σ_A^2 and σ^2, respectively.

For these, Table 77 shows that, in the ANOVA for the first experiment:

The between-groups mean square s_B^2 estimates $\sigma^2 + n\sigma_A^2$.
The within-groups mean square s^2 estimates σ^2.

Hence, by subtraction,

$$s_B^2 - s^2 \text{ estimates } (\sigma^2 + n\sigma_A^2) - \sigma^2 = n\sigma_A^2.$$

so that

$$\hat{\sigma}_A^2 = \frac{s_B^2 - s^2}{n} \quad \text{estimates } \sigma_A^2 \tag{82.2}$$

The numerical values of $\hat{\sigma}_A^2$ and $\hat{\sigma}^2 = s^2$ can now be substituted in (82.1) to estimate the variance of the mean for alternative cost-feasible choices of a_2 and n_2. Since a_2 is a divisor of both variance components in (82.1), it will generally be better, for precision, to increase the number of groups rather than to increase the number of observations per group, especially if $\sigma_A^2 \gg \sigma^2$.

Note: The whole procedure assumes that the population variances do not change appreciably from those estimated by s^2 and $\hat{\sigma}_A^2$ above.

Relative efficiency: The reciprocal of variance is a measure of precision. The relative efficiency of an experiment with a_2 groups and n_2 observations per group as compared with the original investigation can accordingly be estimated as the percentage:

$$\text{Relative efficiency} = \frac{100(\text{precision with } a_2 \text{ and } n_2)}{\text{precision with } a \text{ and } n}$$

$$= \frac{100\{(\hat{\sigma}_A^2/a) + (\hat{\sigma}^2/an)\}}{\{(\hat{\sigma}_A^2/a_2) + (\hat{\sigma}^2/a_2 n_2)\}\%} \tag{82.3}$$

where the numerator is just $100 s_B^2/an$.

Example: From the midge count estimation ANOVA, Table 81.2, the estimates $\hat{\sigma}^2$ and $\hat{\sigma}_A^2$ are

$$\hat{\sigma}^2 = s^2 = 4.8631 \approx 4.9$$

and, from (82.2),

$$\hat{\sigma}_A^2 = \frac{30.6610 - 4.8631}{4} \approx 6.4$$

Table 82. Estimated Variances of $\bar{y}_{..}$ and 95% Confidence Interval Widths

a_2	n_2	$\hat{\sigma}_A^2/a_2$	$\hat{\sigma}^2/n_2 a_2$	$\hat{V}(\bar{y}_{..2})$	95% Confidence Interval Width
3	7	2.13	0.23	2.36	13.2
4	5	1.60	0.25	1.85	8.7
5	4	1.28	0.25	1.53	6.9
10	2	0.64	0.25	0.89	4.3
20	1	0.32	0.25	0.57	3.2

Table 82 shows specimen values of $\hat{V}(\bar{y}_{..2})$, from (82.1), and the widths of the corresponding 95% confidence intervals for μ, for alternative choices of a_2 groups with n_2 replicate observations per group. If, for example, cost considerations allow the choice of $a_2 = 10$ and $n_2 = 2$, the relative efficiency of this with respect to the original investigation is estimated from (82.3) as $100(30.6610/12)/0.89 = 287\%$.

§83. COMPLETELY RANDOMIZED EXPERIMENTS WITH UNEQUAL REPLICATION

Suppose that in a completely randomized experiment there are a groups with n_1 replicate observations in the first group, n_2 in the second group, and so on, and that the unequal replication has not been induced by any of the group effects: if it has, very careful interpretation of findings is required.

The statistical models for y_{ij}, the jth of the n_i observations in the ith group, now become, with the same interpretations as in §72 and §73,

Model I:

$$y_{ij} = \mu + \alpha_i + \varepsilon_{ij}, \qquad \sum \alpha_i = 0, \qquad \varepsilon_{ij} \sim NI(0, \sigma^2)$$
$$j = 1, \ldots, n_i; \quad i = 1, \ldots, a \tag{83.1}$$

Model II:

$$y_{ij} = \mu + A_i + \varepsilon_{ij}, \qquad A_i \sim NI(0, \sigma_A^2), \qquad \varepsilon_{ij} \sim NI(0, \sigma^2) \tag{83.2}$$

with A_i and ε_{ij} independent, for $j = 1, \ldots, n_i; i = 1, \ldots, a$.

Analysis of variance: The ANOVA is the same for both models: With

$$T_i = \sum_{j=1}^{n_i} y_{ij} = \text{the total for the } i\text{th group}$$

$$G = \sum_{1}^{a} T_i = \text{the total of all the observations}$$

$$N = \sum_{i=1}^{a} n_i = \text{the total number of observations,}$$

the ANOVA can be computed as in Table 83.1.

Table 83.1. ANOVA, Completely Randomized Experiment, Unequal
Replication

Variability Source	df	ss	ms
Between-groups	$a-1$	$\sum_{i=1}^{a}\left(\dfrac{T_i^2}{n_i}\right) - \dfrac{G^2}{N}$	$s_B^2 = \dfrac{\text{ss}}{a-1}$
Within-groups	$N-a$	By subtraction	$s^2 = \dfrac{\text{ss}}{N-a}$
Total	$N-1$	$\sum_{i=1}^{a}\sum_{j=1}^{n_i} y_{ij}^2 - \dfrac{G^2}{N}$	

Note: It is sometimes helpful to note that the divisor of a total which is squared in ANOVA calculations is the number of observations contributing to that total. T_i, for example, is a total of n_i observations so that n_i is the divisor for T_i^2; similarly, G^2 is divided by N and $\sum\sum y_{ij}^2$ by 1. The rule obtains to ensure that the population mean of the corresponding mean square is just $V(\varepsilon_{ij}) = \sigma^2$ when H_T is true.

Hypothesis tests: The hypothesis specifications are just those in (72.4), (72.5), and (73.5) for Models I and II, respectively. The test statistic and critical value are

$$F_T = \frac{\text{between-groups ms}}{\text{within-groups ms}} = \frac{s_B^2}{s^2} \quad \text{and} \quad F_\alpha = F(a-1, N-a; \alpha)$$

so that H_T is accepted if $F_T < F_\alpha$ and H_A is accepted if $F_T > F_\alpha$.

Population means of ANOVA mean squares: The results corresponding to those in Table 77 are given in Table 83.2, where

$$n' = \frac{N^2 - \sum n_i^2}{N(a-1)} \tag{83.3}$$

Table 83.2. Population Means of Mean Squares

Variability Source	df	μ(ms) Model I	μ(ms) Model II
Between-groups	$a-1$	$\sigma^2 + \dfrac{\sum_i n_i \alpha_i^2}{a-1}$	$\sigma^2 + n' \sigma_A^2$
Within-groups	$N-a$	σ^2	σ^2

§84. MODEL I CR EXPERIMENTS WITH UNEQUAL REPLICATION: FURTHER ANALYSIS AND EXAMPLE

(i) $F_T < F_\alpha = F(a-1; N-a; \alpha)$: The general remarks in §78 still apply; particular modifications follow.

The $100(1-\alpha)\%$ confidence interval for μ_i is

$$\bar{y}_{i.} \mp t(N-a; \alpha/2) \sqrt{\frac{s^2}{n_i}} \tag{84.1}$$

The $100(1-\alpha)\%$ confidence interval for $(\mu_i - \mu_k)$ is

$$(\bar{y}_{i.} - \bar{y}_{k.}) \mp t(N-a; \alpha/2) \sqrt{s^2 \left(\frac{1}{n_i} + \frac{1}{n_k} \right)} \tag{84.2}$$

The above results follow from (13.5); in both cases s^2 is the ANOVA within-groups mean square.

(ii) $F_T > F_\alpha$: The general remarks in §78 still apply: confidence intervals are calculated from (84.1) and (84.2) and the LSD test is modified so that $H_A: \mu_i \neq \mu_k$ is accepted if

$$|\bar{y}_{i.} - \bar{y}_{k.}| > t(N-a; \alpha/2) \sqrt{s^2 \left(\frac{1}{n_i} + \frac{1}{n_k} \right)} \tag{84.3}$$

Example: Regarding the data in Table 81.1 as referring to a Model I situation, three observations have been omitted to give Table 84.1.

Table 84.1. Transformed Midge Count Data (Modified)

Zone	Replicate y-Values				Total	Mean
1	6.43	3.22	4.40	—	14.05	4.68
2	11.89	10.36	7.24	12.54	42.03	10.51
3	6.95	10.02	—	—	16.97	8.49
			Grand total		73.05	

The ANOVA is computed from Table 83.1 with $n_1 = 3$, $n_2 = 4$, $n_3 = 2$, and $N = 3 + 4 + 2 = 9$. The correction term is $G^2/N = 73.05^2/9 = 592.9225$. Then,

the total sum of squares $= (6.43^2 + \cdots + 10.02^2) - 592.9225 = 85.2246$

and the

between-zones sum of squares

$$= \frac{(14.05)^2}{3} + \frac{(42.03)^2}{4} + \frac{(16.97)^2}{2} - 592.9225 = 58.4990$$

The within-zones sum of squares is obtained by subtraction to complete the ANOVA in Table 84.2.

Based on the model

$$y_{ij} = \mu + \alpha_i + \varepsilon_{ij}, \qquad \alpha_1 + \alpha_2 + \alpha_3 = 0, \qquad \varepsilon_{ij} \sim NI(0, \sigma^2)$$

$$i = 1, 2, 3; \quad j = 1, \ldots, n_i, \quad n_1 = 3, n_2 = 4, n_3 = 2$$

and the specification for testing the equality of the group population means μ_i:

$$H_T: \mu_1 = \mu_2 = \mu_3, \quad H_A: H_T \text{ is false}; \qquad 0.05$$

Table 84.2. ANOVA for Table 84.1 Data

Source	df	ss	ms	F_T
Between-zones	2	58.4990	29.2495	6.57[a]
Within-zones	6	26.7256	4.4542	
Total	8	85.2246		

[a] $0.025 < P < 0.05$.

H_A is accepted because $F_T = 6.57$ exceeds $F(2, 6; 0.05) = 5.14$. For the LSD diagnosis, the means in ascending order are $\bar{y}_1 = 4.68$, $\bar{y}_3 = 8.49$, $\bar{y}_2 = 10.51$. To examine the difference $\bar{y}_3 - \bar{y}_1 = 3.81$, the LSD(0.05), from (84.3) with $t(6; 0.025) = 2.447$, is

$$2.447 \sqrt{4.4542\left(\frac{1}{2}+\frac{1}{3}\right)} = 4.71$$

so that no difference between μ_1 and μ_3 can be inferred. Similarly calculated, the LSD for $\bar{y}_2 - \bar{y}_3 = 2.02$ and $\bar{y}_2 - \bar{y}_1 = 5.83$ are 4.47 and 3.94. We can infer that $\mu_2 = \mu_3$ and $\mu_2 \neq \mu_1$.

The 95% confidence interval for $(\mu_2 - \mu_1)$ is $5.83 \mp 3.94 = 1.89, 9.77$.

§85. MODEL II CR EXPERIMENTS WITH UNEQUAL REPLICATION: FURTHER ANALYSIS AND EXAMPLE

(i) $F_T < F_\alpha = F\{a - 1, N - a; \alpha\}$: The general remarks in §80 apply and, if it is inferred that $\sigma_A^2 = 0$, the observations are regarded as a single sample of $N = \sum_i n_i$ values from just one distribution. Accordingly:
The point estimate of the overall mean μ is

$$\bar{y}_{..} = \frac{\sum_i \sum_{j=1}^{n_i} y_{ij}}{N}$$

The $100(1 - \alpha)\%$ confidence interval for μ is

$$\bar{y}_{..} \mp t(N - 1; \alpha/2) \sqrt{\frac{s_y^2}{N}}$$

where s_y^2 is the mean square in the total row of the ANOVA.

(ii) $F_T > F_\alpha$: The unequal numbers complicate the estimation because the individual group estimates of μ, the $\bar{y}_{i.}$'s, no longer have equal precision. The first of the two estimators, $\hat{\mu}_1$ and $\hat{\mu}_2$, considered here makes some allowance for this; the second does not.

Point estimates of μ:
(a) $\hat{\mu}_1$, the weighted mean of the group means, is given by

$$\hat{\mu}_1 = \frac{\sum_i w_i \bar{y}_{i.}}{\sum_i w_i} \tag{85.1}$$

where the weight $w_i = 1/\hat{V}(\bar{y}_{i.}) = (\hat{\sigma}_A^2 + \hat{\sigma}^2/n_i)^{-1}$ in which, from (82.2) and Table 83.2,

$$\hat{\sigma}^2 = \text{ANOVA within-groups ms}$$

$$\hat{\sigma}_A^2 = \frac{\{\text{between-groups ms} - \text{within-groups ms}\}}{n'} \tag{85.2}$$

with n' from (83.3).

(b) $\hat{\mu}_2$, the unweighted mean of the group means, is given by

$$\hat{\mu}_2 = \frac{\sum_i \bar{y}_{i.}}{a} \tag{85.3}$$

The "goodness" of the $\hat{\mu}_2$ estimate increases the more σ_A^2 exceeds σ^2, the greater are the n_i, and the less are differences between them.

Only approximate confidence intervals are obtainable; a procedure based on $\hat{\mu}_1$ is given in Cochran (1954). For a procedure based on $\hat{\mu}_2$, an approximate $100(1 - \alpha)\%$ confidence interval for μ is

$$\hat{\mu}_2 \mp t(a - 1; \alpha/2)\sqrt{\hat{V}(\hat{\mu}_2)} \tag{85.4}$$

where, because $\hat{\mu}_2$ is the mean of the a $\bar{y}_{i.}$'s,

$$\hat{V}(\hat{\mu}_2) = \frac{\sum \bar{y}_{i.}^2 - a\hat{\mu}_2^2}{a(a - 1)} \tag{85.5}$$

Example: We use the data in Table 84.1.

(a) From (83.3) with $n_1 = 3$, $n_2 = 4$, $n_3 = 2$, and $N = 3 + 4 + 2 = 9$,

$$n' = \frac{\{9^2 - (3^2 + 4^2 + 2^2)\}}{9(2)} = \frac{26}{9}$$

From the ANOVA, Table 84.2, $\hat{\sigma}^2 = 4.4542$ and, using (85.2),

$$\hat{\sigma}_A^2 = \frac{(29.2495 - 4.4542)9}{26} = 8.5830$$

The preliminary computations can then be tabulated as:

$\hat{\sigma}^2/n_i$	$\hat{\sigma}_A^2 + \hat{\sigma}^2/n_i$	$w_i = (\hat{\sigma}_A^2 + \hat{\sigma}^2/n_i)^{-1}$	$\bar{y}_{i.}$
1.4847	10.0677	0.0993	4.68
1.1136	9.6966	0.1031	10.51
2.2271	10.8101	0.0925	8.49

Hence, from (85.1),

$$\hat{\mu}_1 = \frac{0.0993(4.68) + 0.1031(10.51) + 0.0925(8.49)}{0.0994 + 0.1031 + 0.0925} = 7.91$$

(b) For the alternative point estimate $\hat{\mu}_2$, from (85.3),

$$\hat{\mu}_2 = \frac{4.68 + 10.51 + 8.49}{3} = 7.893$$

The two estimates are very close because the differences between these n_i's are relatively small and so the weights in column three above are approximately equal.

For the interval estimate (85.5) gives

$$\hat{V}(\hat{\mu}_2) = \frac{\{4.68^2 + 10.51^2 + 8.49^2 - 3(7.893)^2\}}{6} = 2.924$$

Hence, with $t(2; 0.025) = 4.303$, (85.4) gives a 95% confidence interval for μ (wider than that in §81) as

$$7.893 \mp 4.303\sqrt{2.924} = 0.5, 15.3$$

Planning for precision: The principles and procedures in §82 apply using $\hat{\sigma}_A^2$ calculated from (85.2) instead of (82.2).

§86. THREE-STAGE, RANDOM/RANDOM/RANDOM MODEL, CR INVESTIGATIONS

The preceding Model II estimation procedures can be extended to situations involving more than two sampling stages. Thus, to estimate the midge count over all Iowa marshes (instead of just the one in §81), we could take a random sample of size a from all the Iowa marshes, b zones would then be randomly selected within each marsh, and n independent counts would be made in each of the ab zones. With marshes for the A-stages, zones for the B-stages, and counts for the within-B, within-A stages, this exemplifies the general nested scheme in Fig. 86.

Statistical model: To model y_{ijk}, the kth of the n observations within the jth of the b B-stages within the ith of the a A-stages, we write

$$y_{ijk} = \mu + A_i + B_{ij} + \varepsilon_{ijk} \tag{86.1}$$

with the assumptions that the A_i, B_{ij}, and ε_{ijk} are independent of each

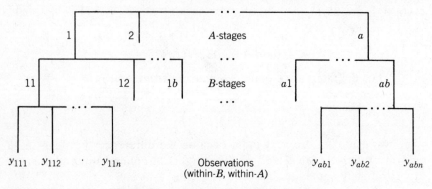

Figure 86. A three-stage nested experiment.

other and

$$A_i \sim NI(0, \sigma_A^2), \qquad B_{ij} \sim NI(0, \sigma_B^2), \qquad \varepsilon_{ijk} \sim NI(0, \sigma^2)$$

for $i = 1, \ldots, a$; $j = 1, \ldots, b$; and $k = 1, \ldots, n$.

It is seen that y_{ijk} differs from its population mean μ by the sum of the three normally distributed deviations A_i, B_{ij}, and ε_{ijk} contributed by the successive stages.

Objectives: The main objective is the point and interval estimation of the overall mean μ in (86.1). Toward this, we test the hypotheses that σ_B^2 and σ_A^2 are zero. If they are not, their estimates can be calculated and used in planning subsequent experiments to estimate μ more precisely.

§87. CR MODEL II EXPERIMENTS; THREE SAMPLING STAGES: NOTATION, AND ANOVA

It is convenient now to systematize the notation as follows.

y_{ijk} = the individual i, j, kth observation, as in (86.1)

$$Y_{ij.} = \sum_{k=1}^{n} y_{ijk} = \text{the total of the } n \text{ observations for the}$$

$$\text{jth B-stage within the ith A-stage}$$

$$Y_{i..} = \sum_{j=1}^{b} Y_{ij.} = \sum_{j=1}^{b} \sum_{k=1}^{n} y_{ijk} = \text{the total of the } bn \text{ observations}$$

$$\text{in the ith A-stage}$$

$$Y_{...} = \sum_{i=1}^{a} Y_{i..} = \sum_{i=1}^{a} \sum_{j=1}^{b} \sum_{k=1}^{n} y_{ijk} = \text{grand total of all the } abn \text{ observations}$$

Corresponding lowercase letters are used to denote the means. Thus, $\bar{y}_{ij.} = Y_{ij.}/n$ is the mean for the i,jth B-stage, $\bar{y}_{i..} = Y_{i..}/bn$ is the ith A-stage mean, and

$$\bar{y}_{...} = \frac{Y_{...}}{abn} = \text{the overall mean, the point estimate of } \mu$$

In essence, there are two ANOVA. The first, Table 87.1, shows how the total sum of squares variability between the y_{ijk}'s is partitioned into sums of squares for variability between and within *all the B-stages* as if there were ab groups with n observations per group in Table 75.1. The second ANOVA, Table 87.2, shows how this sum of squares for the variability between all the B-stages is partitioned into sums of squares to examine the common variability *within* and that *between* the groups of B-stages as if there were a groups with b per group in Table 75.1.

The combined form of the ANOVA is given in Table 87.3. It shows how the total variability, as measured by the sum of squares of the

Table 87.1. ANOVA for ab B-Groups with n Observations per Group

Variability Source	df	ss
Between all B-stages	$ab-1$	$\dfrac{\sum_i \sum_j Y_{ij.}^2}{n} - \dfrac{Y_{...}^2}{abn} = n \sum_i \sum_j (\bar{y}_{ij.} - \bar{y}_{...})^2$
Within all B-stages	$ab(n-1)$	By subtraction $= \sum_i \sum_j \sum_k (y_{ijk} - \bar{y}_{ij.})^2$
Total	$abn-1$	$\sum_i \sum_j \sum_k y_{ijk}^2 - \dfrac{Y_{...}^2}{abn}$

Table 87.2. ANOVA for a Groups with b Subgroups in Each

Variability Source	df	ss
Between A-stages	$a-1$	$\dfrac{\sum_i Y_{i..}^2}{bn} - \dfrac{Y_{...}^2}{abn} = bn \sum_i (\bar{y}_{i..} - \bar{y}_{...})^2$
Between B-stages within A-stages	$a(b-1)$	By subtraction $= n \sum_i \sum_j (\bar{y}_{ij.} - \bar{y}_{i..})^2$
Between all B-stages	$ab-1$	$\dfrac{\sum_i \sum_j Y_{ij.}^2}{n} - \dfrac{Y_{...}^2}{abn} = n \sum_i \sum_j (\bar{y}_{ij.} - \bar{y}_{...})^2$

Table 87.3. ANOVA for Three-Stage Nested Data

Variability Source	df	SS	ms
Between A-stages	$a-1$	$\dfrac{\sum_i Y_{i..}^2}{bn} - \dfrac{Y_{...}^2}{abn}$	$s_1^2 = \dfrac{\text{ss}}{a-1}$
Between B-stages, within A-stages	$a(b-1)$	$\dfrac{\sum_i \sum_j Y_{ij.}^2}{n} - \dfrac{\sum_i Y_{i..}^2}{bn}$	$s_2^2 = \dfrac{\text{ss}}{a(b-1)}$
Within B-stages, within A-stages	$ab(n-1)$	By subtraction	$s^2 = \dfrac{\text{ss}}{ab(n-1)}$
Total	$abn-1$	$\displaystyle\sum_i \sum_j \sum_k y_{ijk}^2 - \dfrac{Y_{...}^2}{abn}$	$s_y^2 = \dfrac{\text{ss}}{(abn-1}$

$(y_{ijk} - \bar{y}_{...})$'s, is partitioned into three component sums of squares, one corresponding to each of the three terms on the right-hand side of the identity

$$(y_{ijk} - \bar{y}_{...}) = (\bar{y}_{i..} - \bar{y}_{...}) + (\bar{y}_{ij.} - \bar{y}_{i..}) + (y_{ijk} - \bar{y}_{ij.})$$

Note: As noted in §83, the divisor of any total which is squared in the Table 87.3 ANOVA sums of squares column is the number of observations contributing to that total. Thus, each $Y_{ij.}$ is a total of n observations so that each $Y_{ij.}^2$, and so $\sum\sum Y_{ij.}^2$, is divided by n.

§88. TESTING THE MEAN SQUARES IN THREE-STAGE MODEL II INVESTIGATIONS

As extensions of the results in §77, it can be shown that the population means of the mean squares in Table 87.3 are as given in Table 88.
If $\sigma_B^2 = 0$ it is seen from Table 88.1 that s_2^2 and s^2, calculated in Table 87.3, are both estimates of σ^2 whereas, if $\sigma_B^2 > 0$, s_2^2 will estimate the quantity $\sigma^2 + n\sigma_B^2$ which exceeds σ^2. Hence, to test

$$H_T: \sigma_B^2 = 0, \quad H_A: \sigma_B^2 > 0; \quad \alpha \qquad (88.1)$$

the test statistic is

$$F_T = \frac{\text{ms between } B\text{-stages, within } A\text{-stages}}{\text{ms within } B\text{-stages, within } A\text{-stages}} = \frac{s_2^2}{s^2} \qquad (88.2)$$

and H_A is indicated if $F_T > F\{a(b-1), ab(n-1); \alpha\}$.

Table 88. Population Means of Mean Squares

Variability Source	df	ms	$\mu[\text{ms}]$
Between A-stages	$a-1$	s_1^2	$\sigma^2 + n\sigma_B^2 + bn\sigma_A^2$
Between B-stages, within A-stages	$a(b-1)$	s_2^2	$\sigma^2 + n\sigma_B^2$
Within B-stages, within A-stages	$ab(n-1)$	s^2	σ^2

For $\sigma_A^2 = 0$ Table 88 shows that both s_1^2 and s_2^2 are estimates of $\sigma^2 + n\sigma_B^2$ whereas, if $\sigma_A^2 > 0$ s_1^2, now estimating $\sigma^2 + n\sigma_B^2 + bn\sigma_A^2$, will probably exceed s_2^2. Hence, to test

$$H_T: \sigma_A^2 = 0, \quad H_A: \alpha_A^2 > 0; \quad \alpha \qquad (88.3)$$

the test statistic is

$$F_T = \frac{\text{ms between } A\text{-stages}}{\text{ms between } B\text{-stages, within } A\text{-stages}} = \frac{s_1^2}{s_2^2} \qquad (88.4)$$

and H_A is indicated if $F_T > F\{(a-1), a(b-1); \alpha\}$.

§89. EXAMPLE: A CR MODEL II EXPERIMENT WITH THREE SAMPLING STAGES

Investigation context and data: Suppose that the (fictitious) data in Table 89.1 are the observed $y = \sqrt{\text{count} + 3/8}$ values for $n = 2$ independent counts obtained in each of $b = 4$ zones, randomly selected in each of $a = 3$ marshes which themselves are regarded as constituting a random sample from a population of such marshes.

Statistical model and notation identifications: With y_{ijk} as the kth of the $n = 2$ count variates in each of the jth of the $b = 4$ zones (the B-stages) within the ith of the $a = 3$ marshes (the A-stages), the model and the assumptions are those in (86.1). Thus,

$$y_{111} = 4.8, \quad y_{112} = 7.4, \quad \dots, \quad y_{341} = 8.6, \quad y_{342} = 4.6$$

The numbers in parentheses in Table 89.1 are marsh-zone totals, for example,

$$Y_{11.} = 4.8 + 7.4 = 12.2, \quad \dots, \quad Y_{34.} = 8.6 + 4.6 = 13.2$$

Table 89.1 Data from a Three-Stage Model II CR Experiment

Marsh	Zone ($b = 1, 2, 3, 4$)				Marsh
$a = 1, 2, 3$	1	2	3	4	Total
1	4.8, 7.4 (12.2)	4.7, 2.9 (7.6)	11.0, 16.1 (27.1)	7.4, 9.7 (17.1)	$64.0 = Y_{1..}$
2	14.5, 20.2 (34.7)	20.1, 14.5 (34.6)	15.6, 14.0 (29.6)	6.5, 7.8 (14.3)	$113.2 = Y_{2..}$
3	5.8, 5.3 (11.1)	6.3, 12.0 (18.3)	1.7, 2.8 (4.5)	8.6, 4.6 (13.2)	$47.1 = Y_{3..}$
				Grand total	$224.3 = Y_{...}$

The first marsh total, for example, is

$$Y_{1..} = 12.2 + 7.6 + 27.1 + 17.1 = 64.0$$

Analysis of variance: The calculations for the ANOVA are

Correction term = $Y^2_{...}/abn = 224.3^2/(3)(4)(2) = 2096.2704$

$\sum \sum \sum y^2_{ijk} = 4.8^2 + 7.4^2 + \cdots + 8.6^2 + 4.6^2 \qquad = 2763.6300$

Total ss $\qquad\qquad\qquad\qquad = 2763.6300 - 2096.2704 = 667.3596$

$\sum Y^2_{i..}/bn = (64.0^2 + 113.2^2 + 47.1^2)/8 \qquad = 2391.0812$

Between marshes ss $\qquad\qquad = 2391.0812 - 2096.2704 = 294.810$

$\dfrac{1}{n} \sum \sum Y^2_{ij.} = \dfrac{1}{2}(12.2^2 + 7.6^2 + \cdots + 13.2^2) \qquad = 2683.9550$

Between zones within marshes ss $= 2683.9550 - 2391.0812 = 292.8738$

With the final subtraction for within-zones, within-marshes, the ANOVA may then be completed as in Table 89.2.

Hypothesis testing: The B-stages are zones with variance σ^2_B. To test

$$H_T: \sigma^2_B = 0, \quad H_A: \sigma^2_B > 0; \quad 0.05$$

Table 89.2. ANOVA for Table 89.1

Source	df	ss	ms	F_T
Between-marshes	2	294.8108	147.4054	4.53[a]
Between-zones, within-marshes	9	292.8738	32.5415	4.86[a]
Within-zones, within-marshes	12	79.6750	6.6396	
Total	23	667.3596		

[a] $P < 0.05$.

the test statistic (88.2) is $F_T = 32.5415/6.6996 = 4.86$ which exceeds $F_\alpha = F(9, 12; 0.05) = 2.80$.

The between A-stages variance is σ_A^2, the variance between marshes. To test

$$H_T: \sigma_A^2 = 0, \quad H_A: \sigma_A^2 > 0; \quad 0.05$$

the test statistic (88.4) is $147.4054/32.5415 = 4.53$ which exceeds $F_\alpha = F(2, 9; 0.05) = 4.26$.

Since the test statistics exceed their critical values in both cases, estimation and planning procedures need to take account of nonzero variability components between marshes and between zones within marshes.

§90. THREE-STAGE MODEL II INVESTIGATIONS: ESTIMATING THE MEAN

A primary objective of a Model II investigation is to estimate the overall mean of the population sampled in the first stage—the parameter μ in (86.1). For this, since the A_i, B_{ij}, and ε_{ijk} all have zero population mean,

$$\textit{The point estimate of } \mu \textit{ is } \bar{y}_{...} = \frac{\sum_i \sum_j \sum_k y_{ijk}}{abn}$$

Interval estimation: Interval estimates differ according to the inferences made about σ_A^2 and σ_B^2. Remembering that, because of Type II error possibilities, the fact that $F_T < F_\alpha$ does not guarantee the truth of any test hypothesis, four cases are noted:

(i) $\sigma_A^2 > 0$, $\sigma_B^2 > 0$. From (86.1) the model for $\bar{y}_{...}$ is, in this case,

$$\bar{y}_{...} = \mu + \bar{A}_. + \bar{B}_{..} + \bar{\varepsilon}_{...} \tag{90.1}$$

where $\bar{A}_.$, $\bar{B}_{..}$, and $\bar{\varepsilon}_{...}$ are the means of a, ab, and abn variates with respective variances σ_A^2, σ_B^2, and σ^2. Hence,

$$V(\bar{y}_{...}) = \frac{\sigma_A^2}{a} + \frac{\sigma_B^2}{ab} + \frac{\sigma^2}{abn} = \frac{\sigma^2 + n\sigma_B^2 + bn\sigma_A^2}{abn} \tag{90.2}$$

so that, from Table 88.1,

$$V(\bar{y}_{...}) = \frac{\{\text{population mean of the between } A\text{-stages ms}\}}{abn}$$

and, accordingly, $V(\bar{y}_{...})$ can be estimated as

$$\hat{V}(\bar{y}_{...}) = \frac{\{\text{ANOVA between } A\text{-stages ms}\}}{abn} = \frac{s_1^2}{abn}$$

from Table 87.3. Then, since s_1^2 has $(a-1)$ degrees of freedom:

The $100(1-\alpha)\%$ *confidence interval for* μ *is* $\bar{y}_{...} \mp t(a-1; \alpha/2)\sqrt{\dfrac{s_1^2}{abn}}$

$$\tag{90.3}$$

Example: Continuing the example from §89, the point estimate for μ is

$$\hat{\mu} = \bar{y}_{...} = \frac{224.3}{24} = 9.35$$

and (90.3) gives the 95% confidence interval

$$9.35 \mp 4.303\sqrt{\frac{147.4054}{24}} = -1.3, 20.0$$

taken as from 0 to 20 because μ cannot be negative in this example.

(ii) $\sigma_A^2 = 0$, $\sigma_B^2 > 0$. If $\sigma_A^2 = 0$ is the correct inference from the test in §88, the A_i quantities in (86.1) are zero and (90.1) reduces to

$$\bar{y}_{...} = \mu + \bar{B}_{..} + \bar{\varepsilon}_{...}$$

giving

$$V(\bar{y}_{...}) = \frac{\sigma_B^2}{ab} + \frac{\sigma^2}{abn} = \frac{\sigma^2 + n\sigma_B^2}{abn} \tag{90.4}$$

If $\sigma_A^2 = 0$, Table 88.1 shows that both s_1^2 and s_2^2 estimate $\sigma^2 + n\sigma_B^2$ so that the pooled estimate, with $(ab - 1)$ degrees of freedom, is

$$s_3^2 = \frac{a(b-1)s_2^2 + (a-1)s_1^2}{ab-1} = \frac{(\sum\sum Y_{ij.}^2/n) - (Y_{...}^2/abn)}{ab-1} \qquad (90.5)$$

With s_3^2 from (90.5) and $t_{\alpha/2} = t(ab - 1; \alpha/2)$ the $100(1 - \alpha)\%$ confidence interval for μ is

$$\bar{y}_{...} \mp t_{\alpha/2}\sqrt{s_3^2/abn} \qquad (90.6)$$

(iii) $\sigma_A^2 > 0$, $\sigma_B^2 = 0$. In this case, from (90.1) and (90.2),

$$\bar{y}_{...} = \mu + \bar{A}_. + \bar{\varepsilon}_{...}$$

$$V(\bar{y}_{...}) = \frac{\sigma_A^2}{a} + \frac{\sigma^2}{abn} = \frac{\sigma^2 + bn\sigma_A^2}{abn} \qquad (90.7)$$

Table 88.1 shows that s_1^2 is the appropriate estimate of $\sigma^2 + bn\sigma_A^2$ so that the $100(1 - \alpha)\%$ confidence interval for μ is, with $t_{\alpha/2} = t(a - 1; \alpha/2)$,

$$\bar{y}_{...} \mp t_{\alpha/2}\sqrt{\frac{s_1^2}{abn}} \qquad (90.8)$$

(iv) $\sigma_A^2 = \sigma_B^2 = 0$. If both σ_A^2 and σ_B^2 are zero we simply have an unstructured sample of abn observations, $y_{ijk} \sim NI(\mu, \sigma^2)$, so that

$$V(\bar{y}_{...}) = \frac{\sigma^2}{abn} \qquad (90.9)$$

The estimate of σ^2 is just s_y^2 in the total row of Table 87.3 and the $100(1 - \alpha)\%$ confidence interval is

$$\bar{y}_{...} \mp t(abn - 1; \alpha/2)\sqrt{\frac{s_y^2}{abn}} \qquad (90.10)$$

§91. THREE-STAGE MODEL II CR INVESTIGATIONS: PLANNING FOR PRECISION

As in §82, components of variance estimated from the ANOVA can be inserted into the appropriate formula for $\hat{V}(\bar{y}_{...})$ to estimate the variance of the point estimate of μ for alternative values of a, b, and n in subsequent experiments. The appropriate formula for $\hat{V}(\bar{y}_{...})$ will be one of those in §90 according to the inferences made about σ_A^2 and σ_B^2. Of these, the $\sigma_A^2 > 0$, $\sigma_B^2 > 0$ is probably the most common and important.

For this case, if $\sigma_A^2 > 0$ and $\sigma_B^2 > 0$ and a_2, b_2, and n_2 are the first, second, and third stage numbers used in a second experiment, the variance $\hat{V}(\bar{y}_{..})_2$ of the new point estimate is estimated from (90.2) as

$$\hat{V}(\bar{y}_{...})_2 = \frac{\hat{\sigma}_A^2}{a_2} + \frac{\hat{\sigma}_B^2}{a_2 b_2} + \frac{\hat{\sigma}^2}{a_2 b_2 n_2} \tag{91.1}$$

From Table 88.1, the required estimates of the variance components are

$$\hat{\sigma}_A^2 = \frac{s_1^2 - s_2^2}{bn}, \qquad \hat{\sigma}_B^2 = \frac{s_2^2 - s^2}{n}, \qquad \hat{\sigma}^2 = s^2 \tag{91.2}$$

Example: (We continue the example from §89 and §90, wherein the wide confidence interval and the unrealistic value for its lower limit do indicate an imprecise experiment.) Toward planning an improvement we find from (91.2) and the ANOVA

$$\hat{\sigma}_A^2 = \frac{147.4054 - 32.5415}{8} = 14.4, \qquad \hat{\sigma}_B^2 = \frac{32.5415 - 6.6396}{2} = 13.0$$

$$\hat{\sigma}^2 = s^2 = 6.6$$

Hence, for example, if we next sample $a_2 = 6$ marshes, $b_2 = 2$ zones in each and make $n_2 = 2$ counts per zone, (91.1) offers the prospect of obtaining

$$\hat{V}(\bar{y}_{...})_2 = \frac{14.4}{6} + \frac{13.0}{12} + \frac{6.6}{24} = 3.8$$

with a 95% confidence interval semiwidth of $t(5; 0.025)\sqrt{3.8} = 5.0$ both representing appreciable reductions from the previous values $\hat{V}(\bar{y}_{...}) = 147.4054/24 = 6.1$ and the interval semiwidth $4.303\sqrt{6.1} = 10.6$. In practice, of course, the potential improvement would need to be assessed against the increased intermarsh travel costs—precision must be paid for!

§92. THREE-STAGE, FIXED/RANDOM/RANDOM MODEL, CR INVESTIGATIONS

In many mixed model nested investigations (i) the elements in the first stage are fixed quantities instead of being random variates, and (ii) the elements in the second, third, ... stages are random variates—obtained by sampling the elements of the preceding stages.

Because of (i), the objective is to compare the individual population means of the first-stage quantities (not to estimate an overall mean).

Example: The Table 89.1 data would refer to a mixed model situation if, instead of being a random sample from a population of marshes, the three marshes were *deliberately selected* for study, for example, to include specific vegetation types and locations.

Statistical model: If y_{ijk} is the kth of the n third-stage observations on the jth of the b B-stage elements, randomly sampled from the ith of the a A-stage elements, the model for y_{ijk} is

$$y_{ijk} = \mu + \alpha_i + B_{ij} + \varepsilon_{ijk} \tag{92.1}$$

with the assumptions that the B_{ij} and ε_{ijk} are independent variates and

$$\sum \alpha_i = 0, \qquad B_{ij} \sim NI(0, \sigma_B^2), \qquad \varepsilon_{ijk} \sim NI(0, \sigma^2)$$

for $i = 1, \ldots, a; j = 1, \ldots, b; k = 1, \ldots, n$

Comparison with (86.1) shows that α_i replaces A_i with $\sum \alpha_i = 0$ instead of the assumption $A_i \sim NI(0, \sigma_A^2)$ and, as in the two-stage fixed model, the population means of interest are again

$$\mu_i = \mu + \alpha_i, \qquad i = 1, \ldots, a \tag{92.2}$$

Analysis of variance: The notation for the data and the analysis of variance computations are exactly as given in §87 and exemplified in §89. The population means of ANOVA mean squares are now those shown in Table 92.

Hypothesis test (1), for the second-stage variance σ_B^2: Table 92 shows that $\mu(s_2^2) > \mu(s^2)$ if $\sigma_B^2 > 0$ and $\mu(s_2^2) = \mu(s^2) = \sigma^2$ if $\sigma_B^2 = 0$. For $H_T: \sigma_B^2 = 0$, $H_A: \sigma_B^2 > 0$; α therefore, H_A is accepted if

$$F_T = \frac{s_2^2}{s^2} > F\{a(b-1). \ ab(n-1); \alpha\}$$

Table 92. Population Mean Squares in the Three-Stage Mixed Model ANOVA

Variability Source	df	ms	$\mu[\text{ms}]$
Between A-stages	$a-1$	s_1^2	$\sigma^2 + n\sigma_B^2 + \dfrac{bn}{a-1}\sum \alpha_i^2$
Between B-stages within A-stages	$a(b-1)$	s_2^2	$\sigma^2 + n\sigma_B^2$
Within B-stages within A-stages	$ab(n-1)$	s^2	σ^2

Hypothesis test (2), for differences between group population means:
(i) $\sigma_B^2 = 0$. If it is accepted that $\sigma_B^2 = 0$, the second-stage differences are uninfluential, the investigation reduces to a simple Model I experiment with a groups and bn observations per group so that, *mutatis mutandis*, the procedures in §78 are applicable.

(ii) $\sigma_B^2 > 0$: If there are no differences between the a fixed effects, all the μ_i's will be the same, all the α_i's will be zero, and so $\sum \alpha_i^2 = 0$. From Table 92 the between A-stages mean square will then estimate the same population quantity $(\sigma^2 + n\sigma_B^2)$ as the between B-stages within A-stages mean square. Hence, to test

$$H_T: \mu_1 = \cdots = \mu_a = \mu, \quad H_A: \text{the } \mu_i\text{'s are not all equal}; \quad \alpha$$

or, equivalently,

$$H_T: \alpha_1 = \cdots = \alpha_a = 0, \quad H_A: H_T \text{ is false}; \quad \alpha \qquad (92.3)$$

the test statistic is obtained from the ANOVA as

$$F_T = \frac{\text{ms between } A\text{-stages}}{\text{ms between } B\text{-stages within } A\text{-stages}} = \frac{s_1^2}{s_2^2} \qquad (92.4)$$

and H_A is accepted if $F_T > F_\alpha = F\{a - 1, a(b - 1); \alpha\}$.

LSD diagnosis: If H_A is accepted, the LSD procedure can be used to examine differences between pairs of means. From (92.1)

$$\bar{y}_{i..} = \frac{\sum \sum (\mu + \alpha_i + B_{ij} + e_{ijk})}{bn} = \mu + \alpha_i + \bar{B}_{i.} + \bar{\varepsilon}_{i..}$$

so that

$$V(\bar{y}_{i..}) = \frac{\sigma_B^2}{b} + \frac{\sigma^2}{bn} = \frac{\sigma^2 + n\sigma_b^2}{bn} \qquad (92.5)$$

Since they are independent, the variance of the difference between two such means is $2(\sigma^2 + n\sigma_B^2)/bn$. From Table 92 the estimate of $(\sigma^2 + n\sigma_B^2)$ is the between B-stages within A-stages mean square s_2^2, with $a(b-1)$ degrees of freedom. For the least significant difference procedure, means are tested in pairs and if any two, $\bar{y}_{i..}$ and $\bar{y}_{h..}$ say, are such that

$$|\bar{y}_{i..} - \bar{y}_{h..}| > t\{a(b - 1); \alpha/2\} \sqrt{\frac{2s_2^2}{bn}} \qquad (92.6)$$

the right-hand side quantity being the LSD, we infer that the corresponding population means, μ_i and μ_h, are different.

Estimations: The point estimate of the group mean, $\mu_i = \mu + \sigma_i$, is

$$\bar{y}_{i..} = \frac{\text{the } i\text{th group total}}{bn}$$

The $100(1 - \alpha)\%$ confidence interval for μ_i is, in consequence of (92.5),

$$\bar{y}_{i..} \mp t\{a(b-1); \alpha/2\}\sqrt{\frac{s_2^2}{bn}} \tag{92.7}$$

The $100(1 - \alpha)\%$ confidence interval for $(\mu_i - \mu_h)$ is

$$(\bar{y}_{i..} - \bar{y}_{h..}) \mp t\{a(b-1); \alpha/2\}\sqrt{\frac{2s_2^2}{bn}} \tag{92.8}$$

the right-hand side being the LSD in (92.6).

Planning for precision: For precise estimations and comparisons, the individual group variance, $V(\bar{y}_{i..})$ in (92.5), should be small. If $(\bar{y}_{i..})_2$ is the ith group mean in a second experiment using b_2 second-stage samples per group and n_2 observations per second-stage sample, we can anticipate

$$\hat{V}(\bar{y}_{i..2}) = \frac{\hat{\sigma}_B^2}{b_2} + \frac{\hat{\sigma}^2}{b_2 n_2} \tag{92.9}$$

into which we substitute, as an outcome from the ANOVA, the estimates

$$\hat{\sigma}_B^2 = \frac{s_2^2 - s^2}{n} \quad \text{and} \quad \hat{\sigma}^2 = s^2 \tag{92.10}$$

for calculations with alternative choices of b_2 and n_2.

§93. EXAMPLE: A THREE-STAGE MIXED MODEL CR INVESTIGATION

Data: As part of a study on artificially induced lactation (Benson et al., 1955), the mean lobule alveolar counts for a = three goats, were compared by taking $b = 4$ random slices from formalin-fixed half udders and making independent counts on $n = 4$ sections for each slice. The data are given in Table 93.1 and the ANOVA, computed exactly as in Tables 87.3 and 89.2, is in Table 93.2.

Statistical model and assumptions: The individual slice-section counts, $y_{111} = 95, \ldots, y_{344} = 139$, are those in sets of $n = 4$ in the 12 cells of

Table 93.1. Lobule Alveolar Counts on Goat Mammary Gland Tissue

Goat	Slice				Total
	1	2	3	4	
467	95, 103 78, 91 (367)	138, 119 104, 129 (490)	138, 147 100, 131 (516)	105, 136 135, 143 (519)	$1892 = Y_{1..}$
469	164, 179 181, 183 (707)	180, 179 119, 130 (608)	138, 171 122, 135 (566)	151, 170 151, 149 (621)	$2502 = Y_{2..}$
484	119, 121 123, 142 (505)	146, 149 162, 153 (610)	127, 154 150, 155 (586)	171, 163 160, 139 (633)	$2334 = Y_{3..}$
				Grand total	$6728 = Y_{...}$

Table 93.1. From (92.1), the model for y_{ijk}, the count for the kth section of the jth slice for the ith goat is

$$y_{ijk} = \mu + \alpha_i + B_{ij} + \varepsilon_{ijk} \qquad (93.1)$$

so that $\mu_1 = \mu + \alpha_1$, $\mu_2 = \mu + \alpha_2$, and $\mu_3 = \mu + \alpha_3$ are the population mean counts for goats 467, 469, and 484, respectively. For the assumptions we take

$$\sum \alpha_i = 0, \qquad B_{ij} \sim NI(0, \sigma_B^2), \qquad \varepsilon_{ijk} \sim NI(0, \sigma^2)$$

Table 93.2. ANOVA for Mammary Gland Count Data ($a = 3$, $b = 4$, $n = 4$)

Source	df	ss	ms	F_T
Between-goats	2	12410.17	6205.09	6.32[a]
Between-slices, within-goats	9	8830.00	981.11	3.70[b]
Between-sections, within-slices	36	9540.50	265.01	
Total	47	30780.67		

[a] $P < 0.025$.
[b] $P < 0.005$.

with B_{ij} and ε_{ijk} independent. The independence assumptions were met in this experiment and, for present purposes, it is assumed that the normality and homogeneous variability specifications in (92.1) are also satisfied. The analysis then proceeds as follows:

Hypothesis test (1) for σ_B^2: The test statistic for $H_T: \sigma_B^2 = 0$, $H_A: \sigma_B^2 > 0$ is

$$F_T = \frac{\text{ms between-slices, within-goats}}{\text{ms between-sections within-slices}} = 3.70$$

as in Table 93.2 and, since $3.70 > F(9, 36; 0.005) \approx 3.25$, we can infer that the slice to slice variance σ_B^2 is not zero and not negligible in planning for precision.

Hypothesis test (2) for differences between goat population means: Inferences about possible goat to goat differences are drawn from

$$F_T = \frac{\text{ms between-goats}}{\text{ms between-slices, within-goats}} = 6.32$$

which exceeds $F(2, 9; 0.025) = 5.715$ to indicate some goat differences and invite LSD diagnosis.

LSD Diagnosis: The goat means $\bar{y}_{i..} = Y_{i..}/16$, arranged in ascending order, are

Goat	467	484	469
Mean	118.25	145.88	156.38

Taking $\alpha = 0.05$ and $t(9; 0.025) = 2.262$, (92.6) gives the LSD(0.05) as $2.262\sqrt{2(981.11)/16} = 25.05$ and since $145.88 - 118.25 > 25.05$ and $156.38 - 145.88 < 25.05$ we infer that the population means for goats 484 and 469 both differed from the mean for goat 467 but were not detected as different from each other.

Estimation: As examples of point and interval estimations, the point estimate of $\mu_{(467)}$ is $\bar{y}_{1..} = 118.25$ and the 95% confidence interval estimate of $\mu_{(467)}$ is, from (92.7),

$$118.25 \mp 2.262 \sqrt{\frac{981.11}{16}} = 100, 136$$

The 95% confidence interval for the difference $\mu_{(469)} - \mu_{(484)}$ is, from (92.8),

$$(156.38 - 145.88) \mp 25.05 \approx -14, 36$$

Planning for precision: For planning an experiment to give narrower confidence intervals for the individual goat means and differences between them, we first find the estimates of the variance components involved. For these, from (92.10) and the ANOVA table,

$$\hat{\sigma}_B^2 = \frac{981.11 - 265.01}{4} = 179.03, \qquad \hat{\sigma}^2 = 265.01$$

Hence, for $n_2 = 2$ counts on each of $b_2 = 8$ slices, involving the same number of counts per goat, (92.9) suggests the achievement in a subsequent experiment, provided the component variances do not change, of the variance estimated by

$$\hat{V}(\bar{y}_{i..})_2 = \frac{179.03}{8} + \frac{265.01}{16} = 39$$

This is appreciably smaller than the value, $981.11/16 = 61$, obtained in the above experiment. The benefits for the estimations and the LSD diagnoses would, of course, have to be assessed against the technical changes required—both the histological procedures and the counting are arduous in such investigations.

Note: The preceding recommendations on planning are not definitive recommendations: only part of the whole data has been analyzed here and the data were obtained when this intriguing area of endocrinological research was very new.

Contrasts for Examining Linear Combinations of Group Means: Multiple Comparisons

§94. INTERGROUP CONTRASTS

It is often required to examine linear combinations of means in addition to simple differences between pairs, as in the §79 example where, to examine the water versus salt effect, the mean for treatment A could be compared with the average of the two salt-treatment means. For this, the difference, termed a contrast or comparison, $\mu_A - \frac{1}{2}(\mu_B + \mu_C)$, could be estimated and tested.

Population contrast: If a groups are investigated in a Model I CR experiment and the group population means are $\mu_1, \ldots, \mu_a$, the linear combination

$$\lambda = c_1 \mu_1 + c_2 \mu_2 + \cdots + c_a \mu_a = \sum_1^a c_i \mu_i \qquad (94.1)$$

is defined to be a *population contrast* if

$$c_1 + c_2 + \cdots + c_a = \sum_1^a c_i = 0 \qquad (94.2)$$

that is, if the sum of the c_i constants or coefficients is zero.

Sample contrast: If $\bar{y}_i = \hat{\mu}_i$ is the sample mean for the ith group, the

171

linear combination of the sample means

$$l = c_1 \bar{y}_1 + c_2 \bar{y} + \cdots + c_a \bar{y}_a = \sum_1^a c_i \bar{y}_i \qquad (94.3)$$

using the same c-coefficients, is the corresponding *sample contrast*.

From the statistical model $y_{ij} = \mu_i + \varepsilon_{ij}$ for a CR experiment it follows that $\bar{y}_i = \mu_i + \bar{\varepsilon}_i$ and hence

$$\mu(l) = \sum_1^a c_i \mu(\bar{y}_i) = \sum_1^a c_i \mu_i = \lambda \qquad (94.4)$$

so that l provides the unbiased sample estimate of λ.

Appropriate choices of the c_i coefficients in (94.1) and (94.3) can be made to generate contrasts of interest in particular investigation contexts.

Contrasts specified before the data are obtained are termed *planned* contrasts; others and especially data-suggested contrasts, are termed *unplanned* contrasts. Different statistical procedures are used to examine the two types.

Note: The following procedures are restricted to investigations with equal numbers of replications per group. Statistical methods texts should be consulted for unequal replication contrast procedures.

Examples:

(i) If $a = 3$, the contrast $\lambda = \mu_1 - \frac{1}{2}(\mu_2 + \mu_3)$ is obtained from (94.1) by taking $c_1 = 1$ and $c_2 = c_3 = -\frac{1}{2}$ and the sample estimate is $l = \bar{y}_1 - \frac{1}{2}(\bar{y}_2 + \bar{y}_3)$ with, in both cases $\sum c_i = 1 - (\frac{1}{2}) - (\frac{1}{2}) = 0$ to satisfy (94.2). Differences between pairs of means are also contrasts, thus, with $c_1 = 1$, $c_2 = -1$, and $c_3 = 0$, (94.1) gives the contrast $(\mu_1 - \mu_2)$ and its estimate is $\bar{y}_1 + (-1)\bar{y}_2 = (\bar{y}_1 - \bar{y}_2)$.

(ii) Taking $c_i = (x_i - \bar{x})/\sum' xx$, the regression coefficient in (48.1) is $b = \sum c_i(y_i - \bar{y}) = \sum c_i y_i - \bar{y} \sum c_i$. Hence, because $\sum c_i = 0$, $b = \sum c_i y_i$ is a sample contrast between the n-observations in the Model I, fixed x-values case.

§95. CONFIDENCE INTERVALS FOR PLANNED POPULATION CONTRASTS

Experimental context: Suppose that for a completely randomized Model I, fixed effects experiment, with a groups and equal numbers of observations n in each group, the contrast of interest and its sample estimate are, respectively,

$$\lambda = \sum c_i \mu_i \quad \text{and} \quad l = \sum c_i \bar{y}_i, \quad \text{with} \sum c_i = 0$$

The $100(1 - \alpha)\%$ confidence interval for $\lambda = \mu(l)$ is then

$$l \mp t_{\alpha/2} \sqrt{\frac{(\sum c_i^2)s^2}{n}} \tag{95.1}$$

where $t_{\alpha/2} = t\{a(n - 1); \alpha/2\}$ and s^2 is the ANOVA within-groups mean square.

Example: From §79, the mean hemolymph sodium concentrations in mosquito larvae were $\bar{y}_A = 87.0$, $\bar{y}_B = 113.5$, and $\bar{y}_C = 100.3$ for the three treatments: $A = $ distilled water, $B = 85$ mM NaCl/liter, and $C = 1.7$ mM NaCl/liter. To examine the water versus salt difference, the contrast of interest is

$$\lambda = \mu_A - \tfrac{1}{2}(\mu_B + \mu_C)$$

from (94.1) with $c_1 = 1$, $c_2 = c_3 = -\tfrac{1}{2}$. Hence, from (94.3), λ is estimated as

$$l = \bar{y}_A - \frac{\bar{y}_B + \bar{y}_C}{2} = 87.0 - \frac{113.5 + 100.3}{2} = -19.9 \text{ Na units}$$

There were $n = 6$ observations per group and $s^2 = 45.25$, 15 df, from the ANOVA (Table 79.2) so that, for a 95% confidence interval, $t_{\alpha/2} = t(15, 0.025) = 2.131$. Next, $\sum c_i^2 = 1 + \tfrac{1}{4} + \tfrac{1}{4} = 1.5$ so that the 95% confidence interval for λ is obtained from (95.1) as

$$-19.9 \mp 2.131 \sqrt{\frac{(1.5)(45.25)}{6}} = -27.1, -12.7 \text{ Na units}$$

The interval excludes zero to indicate H_A in the hypothesis specification:

$$H_T: \mu_A - \tfrac{1}{2}(\mu_B + \mu_C) = 0, \quad H_A: \mu_A - \tfrac{1}{2}(\mu_B + \mu_C) \neq 0; \quad \alpha = 0.05$$

The result (95.1) flows from Rules 1 and 2 in §6 and §7; see also §8 Examples (iii) and (iv). In the present context, for the sample contrast $l = \sum c_i \bar{y}_i$, we have

$$V(l) = V\left(\sum_i c_i \bar{y}_i\right) = \sum_i V(c_i \bar{y}_i)$$

by Rule 2 because the sample means $\bar{y}_{i.}$ are independent. Also, from

(6.2) and (8.2), with $\sigma^2 = V(\varepsilon_{ij})$ in the CR model (72.2),

$$V(c_i \bar{y}_i) = c_i^2 V(\bar{y}_i) = \frac{c_i^2 \sigma^2}{n}$$

so that, altogether,

$$V(l) = \frac{\sum_i c_i^2 \sigma^2}{n} = \left(\sum_i c_i^2\right)\left(\frac{\sigma^2}{n}\right)$$

The estimate of the standard deviation of l is therefore $\sqrt{(\sum c_i^2)(s^2/n)}$, where s^2 is the ANOVA within-groups mean square with $a(n-1)$ degrees of freedom. Taking λ as the parameter θ and $c = \sum c_i^2/n$ in (13.5) then gives the interval (95.1).

Note: Strictly, the $100(1-\alpha)\%$ confidence for the interval (95.1) applies for a single contrast only. If the confidence $100(1-\alpha)\%$ is required for several contrasts, "that λ_1 lies in its interval *and* λ_2 is in its interval *and* $\lambda_3 \ldots$," a modified formula for calculating the intervals is applicable; see Miller (1981).

§96. HYPOTHESIS TESTING FOR PLANNED CONTRASTS: CONTRASTS USING TOTALS

Suppose that, with suitably chosen c-coefficients, the contrast of interest is $\lambda = \sum c_i \mu_i$, $\sum c_i = 0$, and we wish to examine the hypothesis specification

$$H_T: \lambda = 0, \quad H_A: \lambda \neq 0; \quad \alpha \tag{96.1}$$

A two-sided H_A is taken here for example purposes; in practice H_A will be one or two-sided according to the particular experimental context.

The estimate of λ is the sample contrast $l = \sum c_i \bar{y}_i$ and, because $\mu(l) = \lambda$, the test statistic, with $\lambda_T = 0$, is

$$t_T = \frac{l-0}{\sqrt{\hat{V}(l)}} = \frac{l}{\sqrt{(\sum c_i^2)s^2/n}} \tag{96.2}$$

The test statistic is a $t\{a(n-1)\}$-variate when H_T is true so that H_T will be accepted if

$$-t\{a(n-1); \alpha/2\} < t_T < t\{a(n-1); \alpha/2\}$$

and H_A will be accepted if $|t_T| > t\{a(n-1); \alpha/2\}$.

Contrasts in terms of means typically have fractions for some of the

c_i-coefficients as in the above $a = 3$ example. Especially in the equal replication case, it is usually more convenient to test contrasts using the treatment group totals and integer values for the c_i-coefficients. For this, let T_i be the total of the n replicate observations in group i so that $T_i = n\bar{y}_i$, let the constant k be the lowest common multiple of the denominators of the nonzero contrast-coefficients, and let $C_i = kc_i$, $i = 1, \ldots, a$, be the new set of, now integral, coefficients in a contrast defined as

$$\Lambda = nk\lambda = nk \sum c_i\mu_i = n \sum (kc_i)\mu_i = n \sum C_i\mu_i \qquad (96.3)$$

Since l is the estimate of λ, the estimate of Λ is $\hat{\Lambda} = L$ where

$$\hat{\Lambda} = L = nkl = nk \sum c_i\bar{y}_i = n \sum C_i\bar{y}_i = \sum C_iT_i \qquad (96.4)$$

From (96.3) whenever $\lambda = 0$ so is Λ and conversely so that $H_T: \lambda = 0$ is equivalent to $H_T: \Lambda = 0$ for which the test statistic corresponding to that in (96.2) will be $L/\sqrt{\hat{V}(L)}$.

Now, since $L = (nk)l$,

$$\hat{V}(L) = (nk)^2\,\hat{V}(l) \qquad (96.5)$$

so that

$$t_T = \frac{L}{\sqrt{\hat{V}(L)}} = \frac{(nk)l}{\sqrt{(nk)^2\,\hat{V}(l)}} = \frac{l}{\sqrt{\hat{V}(l)}} \qquad (96.6)$$

which is exactly the test statistic in (96.2).

The test statistic is accordingly obtained by calculating $L = \sum C_iT_i$ and $\hat{V}(L)$ as

$$\hat{V}(L) = \hat{V}\left(\sum C_iT_i\right) = \sum C_i^2\,\hat{V}(T_i) = \left(n \sum C_i^2\right)s^2 \qquad (96.7)$$

because $\hat{V}(T_i) = ns^2$, from (8.5), to give

$$t_T = \frac{L}{\sqrt{(n \sum C_i^2)s^2}} \qquad (96.8)$$

Examples:

(i) For the contrast $\lambda = \mu_a - \frac{1}{2}(\mu_B + \mu_C)$ in the §95 example, $c_1 = 1$, $c_2 = c_3 = -\frac{1}{2}$ so that multiplication by $k = 2$ gives the integral coefficients $C_1 = 2$, $C_2 = C_3 = -1$. Since $n = 6$ in that example, we multiply by $nk = 12$ to obtain $\Lambda = 12\lambda$ and, as per (96.4), the estimate, with the totals from Table 79.1, is

$$\hat{\Lambda} = L = 6\{2\bar{y}_A - \bar{y}_B - \bar{y}_C\} = 2T_A - T_B - T_C$$
$$= 2(522) - 681 - 602 = -239$$

The variance estimate (96.7) is, with $s^2 = 45.25$, 15 df,

$$\hat{V}(L) = 6(2^2 + 1^2 + 1^2)s^2 = 36(45.25) = 1629$$

so that the test statistic (96.8) is

$$t_T = \frac{-239}{\sqrt{1629}} = -5.92$$

indicating the acceptance of $H_T: \lambda \neq 0$ because $|t_T| = 5.92 > t(15; 0.005) = 2.947$.

(ii) Suppose that in a CR experiment with $a = 5$ treatment-groups and n replicate observations per treatment, it is desired to compare the average of the first two treatments with the average of the last three treatments by examining the hypothesis specification

$$H_T: \lambda = \left\{ \frac{(\mu_1 + \mu_2)}{2} - \frac{(\mu_3 + \mu_4 + \mu_5)}{3} \right\} = 0, \quad H_A: H_T \text{ is false}; \quad \alpha$$

In λ and its estimate $l = (\bar{y}_1 + \bar{y}_2)/2 - (\bar{y}_3 + \bar{y}_4 + \bar{y}_5)/3$, the coefficients are $c_1 = c_2 = \frac{1}{2}$, $c_3 = c_4 = c_5 = -\frac{1}{3}$ so that the lowest common multiple is $k = 6$. The hypothesis can accordingly be tested in terms of the contrast

$$L = 3T_1 + 3T_2 - 2T_3 - 2T_4 - 2T_5$$

for which, from (96.7) with $C_1 = C_2 = 3$, $C_3 = C_4 = C_5 = -2$,

$$\hat{V}(L) = n(3^2 + 3^2 + 2^2 + 2^2 + 2^2)s^2 = 30ns^2$$

and H_A will be accepted if $|L|/\sqrt{30ns^2} > t\{a(n-1); \alpha/2\}$.

Confidence intervals for λ: Even though the preceding convenient procedure is used in hypothesis testing, it is to be noted that the basic quantity of practical interest is the contrast λ, *in terms of the population means*. The L-contrast procedure, however, can be continued to find the $100(1 - \alpha)\%$ confidence interval for λ as

$$\frac{\{L \mp t_{\alpha/2}\sqrt{\hat{V}(L)}\}}{nk} = \frac{\{L \mp t_{\alpha/2}\sqrt{(n \sum C_i^2)s^2}\}}{nk} \qquad (96.9)$$

where $t_{\alpha/2} = t\{a(n-1); \alpha/2\}$. The result (96.9) obtains because the point estimate of λ is $l = L/nk$ and

$$\sqrt{\hat{V}(l)} = \frac{1}{nk}\sqrt{\hat{V}(L)}$$

Example: Continuing from (i) above, the 95% confidence interval for $\lambda = \mu_A - \frac{1}{2}(\mu_B + \mu_C)$ is, with $nk = 12$ and $t(15; 0.025) = 2.131$,

$$\frac{\{-239 \mp 2.131\sqrt{1629}\}}{12} = -27.1, -12.7$$

§97. ANOVA F-TESTS FOR CONTRASTS

If $\lambda = \sum c_i\mu_i$, $\sum c_i = 0$ is a contrast of interest, the hypothesis $H_T: \lambda = 0$ can be examined by comparing a t-test statistic, (96.2) or (96.8), with $t(\alpha)$ or $t(\alpha/2)$ for the one-sided or two-sided H_A cases, respectively. An F-test procedure is convenient in the latter case, that is, for testing

$$H_T: \lambda = 0, \quad H_A: \lambda \neq 0; \quad \alpha \tag{97.1}$$

Recalling that the square of a $t(f)$-variate is an $F(1, f)$-variate, the test statistic t_T in (96.8) will fall in its critical region whenever t_T^2 falls in *its* critical region, that is, exceeds $F\{1, a(n-1); \alpha\}$. Accordingly, for the specification (97.1) H_A will be accepted, only if the test statistic

$$F_T = \frac{L^2}{(n \sum C_i^2)s^2} > F\{1, a(n-1); \alpha\} \tag{97.2}$$

The quantity

$$\frac{L^2}{n \sum C_i^2} \tag{97.3}$$

is known as the one degree of freedom sum of squares for the contrast L so that the test statistic is the ratio of this sum of squares to the residual mean square s^2, obtained in the ANOVA.

Notes: (i) If H_A is one-sided, the test statistic can be computed as $\sqrt{F_T}$ from (97.2) and tested against the appropriate one-tailed critical value from the $t\{a(n-1)\}$-distribution.

(ii) In the two-sided H_A case, with $F_\alpha = F\{1, a(n-1); \alpha\}$, the confidence interval for $\lambda = \Lambda/nk$ can be conveniently computed as

$$\left(\frac{L}{nk}\right)\left\{1 \mp \sqrt{\frac{F_\alpha}{F_T}}\right\} \tag{97.4}$$

Examples:

(i) For the data in the §95 example, the two-sided specification $H_T: \lambda = 0$, $H_A: \lambda \neq 0$, 0.05 can be tested in terms of the contrast

$$L = 2T_A - T_B - T_C = -239$$

for which, with $n = 6$, the sum of squares is

$$\frac{L^2}{6(2^2 + 1^2 + 1^2)} = \frac{L^2}{36} = 1586.69$$

With $s^2 = 45.25$, 15 df, from the ANOVA, the test statistic (97.2) is then

$$F_T = \frac{1586.69}{45.25} = 35.06$$

which, when compared with $F_{0.05} = F(1, 15; 0.05) = 4.54$, indicates H_A. Furthermore, from (97.4), the 95% confidence interval for λ is computable as

$$\left(\frac{-239}{12}\right)\left\{1 \mp \sqrt{\frac{4.54}{35.06}}\right\} = -27.1, -12.7$$

(ii) If the ANOVA (Table 75.1) is obtained for an $a = 2$, two-treatment completely randomized experiment, the hypotheses $H_T: \mu_1 = \mu_2$, $H_A: \mu_1 \neq \mu_2$; α, can be examined by comparing the ANOVA test statistic $F_T = s_B^2/s^2$ with $F_\alpha = F\{1, 2(n-1); \alpha\}$ and F_T is the square of the test statistic (24.11) when $\Delta_T = 0$. In this situation, with $\lambda = \mu_1 - \mu_2$, the equivalent specification

$$H_A: \lambda = 0, \quad H_A: \lambda \neq 0; \quad \alpha$$

is tested using the contrast $L = T_1 - T_2$ for which the sum of squares is $L^2/n(1 + 1) = L^2/2n$, giving the test statistic (97.2) as $F_T = L^2/2ns^2$. It is a simple exercise to show that $L^2/2n$ is exactly equal to the between-groups mean square s_B^2 in Table 75.1.

§98. MUTUALLY ORTHOGONAL CONTRASTS

Suppose that T_i, $i = 1, 2, \ldots, a$, are the independent, individual group totals, of n observations each, in a completely randomized experiment and that two contrasts are constructed from two different sets of coefficients, C_i and D_i, as

$$L_1 = \sum_1^a C_i T_i \quad \text{and} \quad L_2 = \sum_1^a D_i T_i, \quad \text{with} \quad \sum C_i = \sum D_i = 0$$

Then the contrasts L_1 and L_2 are said to be *orthogonal* if the sum of the products of corresponding coefficients is zero, that is, if

$$\sum_1^a C_i D_i = 0 \qquad (98.1)$$

Similarly, in terms of means, if $\bar{y}_i$, $i = 1, \ldots, a$, is the set of a independent means, each of n independent observations, the two contrasts

$$l_1 = \sum c_i \bar{y}_i \quad \text{and} \quad l_2 = \sum d_i \bar{y}_i, \quad \text{with} \quad \sum c_i = \sum d_i = 0$$

are orthogonal if

$$\sum_1^a c_i d_i = 0$$

Example: Suppose that with $a = 3$ groups, four contrasts of practical interest are $(\mu_1 - \mu_2)$, $(\mu_1 + \mu_2)/2 - \mu_3$, $(\mu_1 - \mu_3)$, and $(\mu_1 + \mu_3)/2 - \mu_2$, which can be examined in terms of the sample contrasts among the group totals T_i. The contrasts are

$$L_1 = T_1 - T_2, \quad L_2 = T_1 + T_2 - 2T_2,$$

$$L_3 = (T_1 - T_3), \quad \text{and} \quad L_4 = T_1 - 2T_2 + T_3$$

Then, from the four respective coefficient sets,

$$(1, -1, 0), \quad (1, 1, -2), \quad (1, 0, -1), \quad \text{and} \quad (1, -2, 1)$$

for L_1, L_2, L_3, and L_4, we find, using (98.1), that

L_1 and L_2 are orthogonal because $(1)(1) + (-1)(1) + 0(-2) = 0$.
L_1 and L_3 are not orthogonal because $(1)(1) + (-1)(0) + 0(-1) \neq 0$.
L_3 and L_4 are orthogonal because $(1)(1) + 0(-2) + (-1)(1) = 0$.

It can similarly be checked that no other two of these contrasts are orthogonal, and that the outcomes are the same for the corresponding contrasts in terms of the group means; thus, $l_2 = (\bar{y}_1 + \bar{y}_2)/2 - \bar{y}_3$ is not orthogonal to $l_3 = \bar{y}_1 - \bar{y}_3$ because $(\frac{1}{2})1 + (\frac{1}{2})0 - 1(-1) \neq 0$.

An important property of orthogonal contrasts is that from the a group totals it is possible to form different sets of contrasts, each of which is orthogonal to all the others in its set. Each set, however, can have no more than $(a - 1)$ contrasts.

Example: If $a = 4$, two sets of three mutually orthogonal contrasts are:

(i) $L_1 = T_1 - T_2$, $L_2 = T_1 + T_2 - 2T_3$, $L_3 = T_1 + T_2 + T_3 - 3T_4$.

(ii) $L_1 = T_1 + T_2 - T_3 - T_4$, $L_2 = T_1 - T_2 + T_3 - T_4$,
$L_3 = T_1 - T_2 - T_3 + T_4$.

In practice, contrasts are selected to investigate particular combinations of the population means of interest in investigations. Thus, for the data in §79, the orthogonal contrasts $\lambda_1 = \mu_A - \frac{1}{2}(\mu_B + \mu_C)$ and $\lambda_2 = (\mu_B - \mu_C)$ can be used, respectively, to compare the water with the average of the two salt treatments and the high salt with the low salt treatment.

Orthogonal contrasts are convenient for hypothesis testing because of the fact that, if, in an a-group experiment, $L_1, L_2, \ldots, L_{a-1}$ is a set of mutually orthogonal contrasts, then:

The sum of the $(a - 1)$ sums of squares (97.3) for the individual contrasts is equal to the ANOVA between-groups sum of squares.

Algebraically, if the contrasts are $L_k = \sum_{i=1}^{a} C_{ki}T_i$, $k = 1, 2, \ldots, (a-1)$, then

$$\frac{L_1^2}{n \sum C_{1i}^2} + \frac{L_2^2}{n \sum C_{2i}^2} + \cdots + \frac{L_{a-1}^2}{n \sum C_{a-1i}^2} = \frac{\sum T_i^2}{n} - \frac{G^2}{an} \qquad (98.2)$$

where $G = \sum_1^a T_i$ is the overall total of all the observations and the quantity on the right-hand side is the between-groups sum of squares in the ANOVA, Table 75.1.

If only $r < (a - 1)$ contrasts are of practical interest, the between-groups sum of squares may be partitioned into individual one degree of freedom components, $ss(L_1), \ldots, ss(L_r)$, and the remainder, with $(a - 1 - r)$ degrees of freedom. The effect proportional to any one of the r planned contrasts can then be tested for significance by calculating the test statistic (97.2) for comparison with $F_\alpha = F\{1, a(n - 1); \alpha\}$.

Example: Suppose that the (entirely fictitious) data on blood cholesterol concentrations presented in Table 98.1 were obtained in a CR experiment with $n = 10$ subjects in each of $a = 5$ groups, representing different primary protein sources, and that the total, corrected, ANOVA sum of squares was 8040, 49 df.

Suppose also that the experiment was planned to investigate the four contrasts defined according to the scheme of contrast coefficients, in

Table 98.1. Fictitious Cholesterol Concentration (mg/10 ml) Data

Group	Primary Protein Source	Total	Mean
A	Animal	350	35
B	Poultry, eggs, cheese	280	28
C	Fish	220	22
D	Vegetables, eggs, cheese	240	24
E	Vegetables only	130	13

terms of totals, in the body of Table 98.2 where, for example, the first contrast compares the carnivore and vegetarian groups and the last contrast compares the near-vegetarians with the vegetarians.

The entries in the last column of Table 98.2 are the divisors, $n \sum C_i^2$. In the denominator of (97.3), for example, the divisor to obtain the sum of squares for the first contrast L_1 is $10(2^2 + 2^2 + 2^2 + 3^2 + 3^2) = 300$. We now compute the four ANOVA individual contrast sums of squares. For the first contrast,

$$L_1 = 2(350) + 2(280) + 2(220) - 3(240) - 3(130) = 590$$

giving the ANOVA ss, from (97.3), as

$$ss(L_1) = \frac{L_1^2}{n \sum C_{1i}^2} = \frac{(590)^2}{300} = 1160.3$$

and similarly $ss(L_2) = 190^2/60 = 601.7$, $ss(L_3) = 245.0$, and $ss(L_4) = 605.0$.

Contrasts of practical interest do not necessarily constitute an orthogonal set such as those above which were deliberately selected to illustrate the additive property (98.2). This, with the total ss = 8048, enables the within-groups ss to be obtained by subtraction to complete the ANOVA in Table 98.3.

Table 98.2. Coefficients for Four Orthogonal Contrasts

	A (350)	B (280)	Group (Group Total) C (220)	D (240)	E (130)	Divisor
$\frac{1}{3}(A+B+C)-\frac{1}{2}(D+E)$	2	2	2	-3	-3	300
$\frac{1}{2}(A+B)-C$	1	1	-2	0	0	60
$A-B$	1	-1	0	0	0	20
$D-E$	0	0	0	1	-1	20

Table 98.3. An ANOVA Partitioned into Contrast Sums of Squares

Variability Source	df	ss	ms	F_T
L_1	1	1160.3	1160.3	9.61[a]
L_2	1	601.7	601.7	4.98[b]
L_3	1	245.0	245.0	2.03
L_4	1	605.0	605.0	5.01[b]
Between-groups	4	2612.0		
Within-groups	45	5436.0	120.8	
Total	49	8048.0		

[a] $P < 0.005$.
[b] $P < 0.05$.

The F_T-statistics for testing H_T: $\lambda_i = 0$, H_A: $\lambda_i \neq 0$, $i = 1, \ldots, 4$, with critical regions from the $F(1, 45)$-distribution, are conveniently presented in the last column of the ANOVA: that for λ_1, for example, is the ratio $ss(L_1)/s^2 = 1160.3/120.8$. Were this "real-life" data we could infer that the carnivores had higher cholesterol concentrations than the vegetarians, that the piscivores had lower concentrations than the other carnivores while the no difference between groups A and B and the significant difference between groups D and E could be attributed to the appreciable cholesterol contents of eggs and cheese.

Although contrasts are conveniently tested using sample totals, the quantities of practical interest are point and interval estimates of the effects measured by the λ-contrasts, those between the group population *means*. Thus, the point estimate of the carnivore versus vegetarian effect,

is
$$\lambda_1 = \tfrac{1}{3}(\mu_A + \mu_B + \mu_C) - \tfrac{1}{2}(\mu_D + \mu_E),$$
$$\lambda_1 = l_1 = \tfrac{1}{3}(35 + 28 + 22) - \tfrac{1}{2}(24 + 13) = 9.8 \text{ mg/10 ml}$$

and, since $L_1 = nkl_1$, where $n = 10$ and $k = 2(3) = 6$, the required 95% confidence interval for λ_1 can be calculated from (96.9) or, using (97.4), as $(L_1/nk)\{1 \mp \sqrt{F_\alpha/F_T}\} = (590/60)\{1 \mp \sqrt{4.057/9.61}\} = 3.4, 16.2 \text{ mg/10 ml}$.

§99. MULTIPLE COMPARISON PROCEDURES FOR TESTING UNPLANNED DIFFERENCES

The previous contrast examination procedures are applicable in experiments where the groups are so structured that intergroup contrasts can be specified before data are obtained. In other cases, for example, in some crop variety selection trials, where many unplanned differences arise for testing, multiple comparison procedures are more appropriate. Investigators can choose between several alternative multiple comparison procedures according to the particular need for:

(a) *Sensitivity*. A sensitive procedure has a relatively high probability of detecting true effects but a correspondingly relatively high probability of erroneously declaring null effects to be non-null.

(b) *Conservatism*. A conservative procedure has a relatively low probability of declaring null effects to be non-null but a correspondingly higher probability of not detecting genuinely non-null effects.

As a generalization of the usual Type I error probability to the multiple comparison context, the desired experimentwise error rate is specified. The experimentwise error rate is defined in the conceptual context of a series of experiments, in which there are no differences between the population means, as

$$\text{Experimentwise error rate} = \frac{\text{number of experiments with at least one false inference}}{\text{total number of experiments}}$$

Assessments of the relative merits of several alternative procedures are given in Bancroft (1968), Miller (1977), and Stoline (1981). The LSD method (§78) is the most sensitive procedure; three others described here are, in order of decreasing sensitivity, the Newman–Keuls, Tukey (HSD), and Scheffé methods.

In the special circumstances when several group means are to be compared with one, usually a control group mean, statistical texts should be consulted for the appropriate procedure; see Dunnett (1955, 1964).

§100. DISTRIBUTION OF THE STUDENTIZED RANGE

Suppose that, in a CR experiment, there are a groups with n replications per group and no differences between the group population means so that $\mu_1 = \mu_2 = \cdots = \mu_a$. Suppose also that $\bar{y}_{\min}$ and $\bar{y}_{\max}$ are the smallest and largest of the a sample means and that $s_{\bar{y}} = s/\sqrt{n}$ is the estimated standard deviation of each group mean where s^2, from the ANOVA, has $a(n-1)$ degrees of freedom for CR experiments. Then the Studentized range q is defined as

$$q = \frac{\bar{y}_{\max} - \bar{y}_{\min}}{s_{\bar{y}}} \qquad (100.1)$$

The parameters for q-distributions are the number of means being tested, r say, and the number of degrees of freedom f for s^2. Right tail-area percentage points are given in Table A6. For experimentwise error rate α, the tabulated entries are $q_{\alpha} = q(r, f; \alpha)$ such that

$$P[q > q(r, f; \alpha)] = \alpha, \quad \text{for } \alpha = 0.05, 0.01$$

where α is the experimentwise error rate.

§101. THE NEWMAN–KEULS (NK) PROCEDURE

This procedure can be used to search for significant differences between the means of groups having the same number of replications n per group. The procedure is not used to obtain confidence intervals. In operation, the a sample means, $\bar{y}_1, \ldots, \bar{y}_a$, are first rearranged in ascending order from smallest to largest. The most extreme difference, that between the largest and smallest mean, is tested first. If it is significant, implying a difference between the corresponding population means, the two next extreme differences, (i) that between the largest and the second smallest and (ii) that between the second largest and the smallest, are tested. The procedure is continued, and when the extreme sample means in any subgroup are not found to be significantly different, it is assumed that none of the means within that subgroup are significantly different. To allow for the fact that, when the population means are equal, the mean range of a set of sample means increases with the number of sample means in the subgroup, the critical value changes from stage to stage. For a (sub)group of r ($2 \le r \le a$) sample means, the population means corresponding to the two extreme sample means are inferred to differ if

(the largest mean) − (the smallest mean) $> D_{\alpha}(r) = q(r, f; \alpha)s_{\bar{y}}$ (101.1)

where, for CR experiments, $f = a(n - 1)$.

Example: Suppose that $a = 5$ group means, of $n = 4$ observations each, are 38.3, 27.8, 13.1, 40.5, and 14.5 for groups A, B, C, D, and E, respectively, and that $s^2 = 72.2$ with $f = a(n - 1) = 15$ df, so that $s_{\bar{y}} = s/\sqrt{n} = \sqrt{72.2/4} = 4.25$. Rearranged in ascending order, the means are

Group	C	E	B	A	D
Mean	13.1	14.5	27.8	38.3	40.5

STEP 1: Beginning with $r = a = 5$, Table A6 gives $q(5, 15; 0.05) = 4.37$ so that $D_{0.05}(5) = (4.37)(4.25) = 18.6$. The difference between the two extreme means $(\bar{y}_D - \bar{y}_C)$ is 27.4 which exceeds 18.6, giving the inference that $\mu_D \neq \mu_C$.

STEP 2: The two sets of $r = 4$ means, (C, E, B, A) and (E, B, A, D), are now examined. With $q(4, 15; 0.05) = 4.08$, $D_{0.05}(4) = (4.08)(4.25) = 17.3$ and because both $(\bar{y}_A - \bar{y}_C) = 25.2$ and $(\bar{y}_D - \bar{y}_E) = 26$ exceed 17.3 we infer that $\mu_A \neq \mu_C$ and $\mu_D \neq \mu_E$.

STEP 3: For $r = 3$, $q(3, 15; 0.05) = 3.67$, $D_{0.05}(3) = 15.6$. Then,

   (i) $\bar{y}_D - \bar{y}_B = 12.7 < 15.6$ and no difference between μ_D and μ_B is inferred.

   (ii) $\bar{y}_A - \bar{y}_E = 23.8 > 15.6$ with the inference that $\mu_A \neq \mu_E$.

   (iii) $\bar{y}_B - \bar{y}_C = 14.7 < 15.6$ and no difference between μ_B and μ_C is inferred.

The findings can be summarized by drawing lines, as above, beneath any subgroup of the ordered means in which the extreme members are not significantly different so that population means of groups, such as A and E above, not having a common underline are inferred to be different.

§102. THE TUKEY (HSD) PROCEDURE

HSD abbreviates honestly significant difference for this Tukey procedure which, for experimentwise error rate α, simply uses the one "yardstick" or distance

$$D_\alpha(a) = q(a, f; \alpha)s_{\bar{y}} \tag{102.1}$$

where $f = a(n - 1)$ in CR experiments, to examine *any* of the $\binom{a}{2} = a(a - 1)/2$ differences between pairs of means. Thus, any two means μ_j and μ_k would be inferred to differ if

$$|\bar{y}_j - \bar{y}_k| > D_\alpha(a) \tag{102.2}$$

It is because the yardstick is not sequentially modified as in the NK procedure that the procedure here is the more conservative.

Example: For the data in the preceding section and $\alpha = 0.05$ we find $D_{0.05}(5) = (4.37)(4.25) = 18.6$. Then, beginning with the ordered arrangement

Group	C	E	B	A	D
Mean	13.1	14.5	27.8	38.3	40.5

Since $14.5 - 13.1 = 1.4$ and $27.8 - 13.1 = 14.7$ are both less than 18.6, we infer that $\mu_C = \mu_E = \mu_B$.

Since $38.3 - 13.1 > 18.6$, we infer that $\mu_A \neq \mu_C$ and $\mu_D \neq \mu_C$.

Since $38.3 - 14.5 > 18.6$, we infer that $\mu_A \neq \mu_E$ and $\mu_D \neq \mu_E$.

Since $38.3 - 27.8 < 18.6$ and $40.5 - 27.8 < 18.6$, we infer that $\mu_B = \mu_A = \mu_D$.

In this example the inferences from the NK and Tukey (HSD) procedures are the same, as indicated by the lines beneath the groups of equal means.

§103. THE SCHEFFÉ PROCEDURE

This procedure can be used to test any and any number of contrasts, including extreme ones selected by data inspection, differences between pairs of means based on unequal replication numbers, and contrasts of no discernible practical relevance. Such catholicism leads to the most conservative properties of this procedure in practice, conservatism that nevertheless may be important in some contexts.

In operation, the procedure controls the experimentwise error rate at $\alpha\%$ and gives a probability of *at most* α *that any* individual contrast is erroneously declared significant. For a groups, if $l = \sum_1^a c_i \bar{y}_a$ is any contrast, the hypothesis H_T: $\mu(l) = \lambda = 0$ is rejected if

$$l^2 > (a-1)F_\alpha \hat{V}(l) = (a-1)F_\alpha s^2 \sum \left(\frac{c_i^2}{n_i}\right) \qquad (103.1)$$

where $F_\alpha = F(a-1, f; \alpha)$ and f = number of degrees of freedom for s^2. Correspondingly, the $100(1-\alpha)\%$ confidence interval for $\mu(l) = \lambda$ is

$$l \mp \sqrt{(a-1)F_\alpha s^2 \sum \left(\frac{c_i^2}{n_i}\right)} \qquad (103.2)$$

The difference between two means $(\mu_j - \mu_k)$ is a simple contrast wherein all the c-coefficients are zero except for $c_j = 1$ and $c_k = -1$. In this case (103.1) will indicate a real difference between μ_j and μ_k if

$$|\bar{y}_j - \bar{y}_k| > \sqrt{(a-1)F_\alpha s^2 \left(\frac{1}{n_j} + \frac{1}{n_k}\right)} \qquad (103.3)$$

Examples:

(i) To test differences between pairs of means for the examples in

§101 and §102 where $n = 4$ for all the groups, the right-hand side of (103.3) gives the yardstick

$$\sqrt{(5-1)F(4, 15; 0.05)(72.2)\tfrac{2}{4}} = \sqrt{(3.06)(72.2)2} = 21.0$$

Although it is not generally so, the inferences from the Scheffé and HSD procedures are the same for this data.

(ii) Summary data from the example in §84 with $a = 3$ and $s^2 = 4.4542$, 6 df, are

Zone	1	2	3
Mean ($\bar{y}_i$)	4.68	10.51	8.49
n_i	3	4	2

Suppose that two contrasts of practical interest are $\lambda_1 = \mu_1 - \mu_2$ and $\lambda_2 = \mu_1 - \tfrac{1}{2}(\mu_2 + \mu_3)$. We first find, for $\alpha = 0.05$, $F(2, 6; 0.05) = 5.14$. Then we can infer, using (103.3), that $\mu_1 \neq \mu_2$ because

$$|\bar{y}_1 - \bar{y}_2| = 5.83 > \sqrt{2(5.14)(4.4542)(\tfrac{1}{3} + \tfrac{1}{4})} = 5.17$$

The estimate of λ_2 is $l_2 = \bar{y}_1 - \tfrac{1}{2}(\bar{y}_2 + \bar{y}_3) = -4.82$ and, with $\sum c_i^2/n_i = (\tfrac{1}{3}) + (\tfrac{1}{2})^2(\tfrac{1}{4}) + (\tfrac{1}{2})^2(\tfrac{1}{2}) = \tfrac{25}{48}$, we find that λ_2 just fails to be detected as different from zero because, using (103.1),

$$4.82 < \sqrt{2(5.14)(4.4542)(\tfrac{25}{48})} = 4.88$$

Furthermore, using (103.2), the 95% level confidence intervals for λ_1 and λ_2 are

$$-5.83 \mp 5.17 = (-11.0 - 0.7) \quad \text{and} \quad -4.82 \mp 4.88 = (-9.7, 0.1)$$

giving at least 95% confidence in the *joint* statement that $-11.0 < \mu_1 - \mu_2 < -0.7$ *and* $-9.7 < \mu_1 - \tfrac{1}{2}(\mu_2 + \mu_3) < 0.1$.

Experiments Using Randomized Block and Latin Square Designs

§104. CONTEXTS FOR RANDOMIZED BLOCK (RB) DESIGNS

Just as the procedures in §70ff extend the procedures in §24 for completely randomized (CR) experiments, randomized block (RB) procedures extend the randomized pair (RP) procedures, §23, to allow population means of $a \geq 2$ groups, or treatments, to be compared.

In RB designs, a total number ab of experimental units is divided into b blocks of a units each where, for the precision advantages, the units in each block are as uniform as possible. For RP designs, $a = 2$.

Examples: Nutrition experiments can be carried out using a rat pups selected at random from b litters to enhance precision by exploiting the fact that pup-to-pup variability within litters is usually appreciably less than that between pups in different litters. In crop variety trials, blocks, each consisting of a group of a contiguous plots, are used to exploit local homogeneity in soil characteristics. Wear characteristics of $a = 4$ experimental tire treads might be compared by observing the results of using the four treads, one on each wheel, on $b = 20$ cars.

Randomization: In the simplest case to be considered, each experimental unit within each block receives and provides a response for one, of the a treatments, selected randomly (Table A1). Thus, in the tire tread example, treads would be allocated at random, one to each of the four wheel positions, and different randomizations would be used from one car to another.

Fixed and random effects: Further to the fixed and random contexts defined for CR experiments, either or both of the block and treatment effects may now be fixed or random according to the objectives of the investigation. In mixed models, blocks could be chosen so that their effects constitute a random sample from a population of circumstances to which inferences about a set of fixed treatments are required.

§105. EXAMPLE AND NOTATION

Table 105 shows the hypothetical results of a nutrition experiment comparing the effects of $a = 4$ treatments, A, B, C, D, in each of $b = 3$ litter-mate blocks. The responses y are weight gains in pounds.

The general notation, exemplified for the Table 105 data, is as follows.

y_{ij} = response to the ith treatment in the jth block.

Thus, $y_{41} = y_{D1} = 5$, $\quad y_{32} = y_{C2} = 9$, $\quad y_{23} = y_{B3} = 6$.

T_i = total of the b responses for the ith treatment;

$\bar{y}_{i.}$ = ith treatment mean.

Thus, $T_1 = T_A = 6 + 8 + 4 = 18$; $\quad \bar{y}_{1.} = \dfrac{18}{3} = 6$.

B_j = total of the a responses in the jth block; $\bar{y}_{.j}$ = jth block mean.

Thus, $B_1 = 6 + 6 + 5 + 9 = 26$; $\quad \bar{y}_{.1} = \dfrac{26}{4} = 6.5$.

$$G = \text{overall total} = \sum_i \sum_j y_{ij} = \sum_i T_i = \sum_j B_j; \; \bar{y}_{..} = \text{overall mean}$$
$$= \frac{G}{ab} = \frac{\sum_i \bar{y}_{i.}}{a}.$$

Thus, $G = 6 + 6 + \cdots + 4 + 6 = 78$; $\quad \bar{y}_{..} = \dfrac{78}{12} = 6.5$.

Table 105. Data in a RB Experiment

	Block (Litter or Dam)											
	1				2				3			
Animal number	1	2	3	4	5	6	7	8	9	10	11	12
Treatment	B	A	D	C	A	C	D	B	D	C	A	B
Response (lb)	6	6	5	9	8	9	6	9	4	6	4	6

§106. STATISTICAL MODELS

When block and treatment effects are fixed, y_{ij} is modeled as

$$y_{ij} = \mu + \alpha_i + \beta_j + \varepsilon_{ij} \tag{106.1}$$

with $\sum_i \alpha_i = \sum_j \beta_j = 0$ and $\varepsilon_{ij} \sim NI(0, \sigma^2)$ for $i = 1, \ldots, a$; $j = 1, \ldots, b$. Here

μ, estimated by $\bar{y}_{..}$, represents the overall mean;

α_i, estimated by $(\bar{y}_{i.} - \bar{y}_{..})$, represents the deviation due to the ith treatment effect so that μ_i, estimated by $\bar{y}_{i.}$, is the ith population mean;

β_j, estimated by $(\bar{y}_{.j} - \bar{y}_{..})$, represents the deviation due to the jth block effect;

ε_{ij} is the random variability component.

When treatment effects are random, α_i in (106.1) is replaced by A_i and $\sum_i \alpha_i = 0$ by the assumption $A_i \sim NI(0, \sigma_A^2)$. When block effects are random, β_j is replaced by B_j and $\sum_j \beta_j = 0$ by the assumption $B_j \sim NI(0, \sigma_B^2)$.

Interactions: When differences between treatments are not constant, apart from random measurement variability, but depend on the particular blocks in which the treatment estimates are made, treatment by block interaction terms are needed in the models. No direct test for the presence of such interactions can be made, however, in the common case analyzed here, when each treatment is represented only once in each block in which case any interaction contributions are latent in the ε_{ij}-components.

§107. STATISTICAL ANALYSIS

As the basis for appropriate hypothesis testing, the ANOVA is first completed as in Table 107.1 in accordance with the ANOVA identity:

$$\sum_i \sum_j (y_{ij} - \bar{y}_{..})^2 = a \sum_j (\bar{y}_{.j} - \bar{y}_{..})^2 + b \sum_i (\bar{y}_{i.} - \bar{y}_{..})^2$$

$$+ \sum_i \sum_j (y_{ij} - \bar{y}_{i.} - \bar{y}_{.j} + \bar{y}_{..})^2 \tag{107.1}$$

which can be read from left to right, as

Total, corrected ss = ss(blocks) + ss(treatments) + ss(residual)

In practice, as in Table 107.1, the computations are more conveniently

Table 107.1. ANOVA for a Randomized Block Experiment (a Treatments, b Blocks)

Variability Source	df	ss	ms
Between-blocks	$b-1$	$\dfrac{1}{a}\sum_j B_j^2 - \left(\dfrac{G^2}{ab}\right)$	s_2^2
Between-treatments	$a-1$	$\dfrac{1}{b}\sum_i T_i^2 - \left(\dfrac{G^2}{ab}\right)$	s_1^2
Residual	$(a-1)(b-1)$	By subtraction	s^2
Total	$ab-1$	$\sum_i \sum_j y_{ij}^2 - \left(\dfrac{G^2}{ab}\right)$	

carried out using totals rather than means and the residual sum of squares is obtained by the subtraction,

$$\text{ss(residual)} = \text{ss(total)} - \text{ss(blocks)} - \text{ss(treatments)}$$

The degrees of freedom are additive, because

$$(b-1)+(a-1)+(ab-a-b+1) = ab-1$$

the sums of squares are additive, from (107.1); the mean squares are not additive.

Hypothesis tests:

(i) *Interactions absent:* The test statistic

$$F_T = \frac{(\text{ms between-treatments})}{(\text{ms within-treatments})} = \frac{s_1^2}{s^2}$$

is used to test the hypotheses previously defined for CR experiments, of equal or unequal population means when treatment effects are fixed (§72) and to test whether $\sigma_A^2 = 0$ or not, when treatment effects are random.

(ii) *Interactions present:* The statistic F_T tests the hypothesis $H_T: \alpha_1 = \alpha_2 = \cdots = \alpha_a = 0$ when block effects are random and treatment effects are fixed. When both block and treatment effects are fixed it is not straightforward to interpret just what F_T usefully tests.

Further analyses: When the treatment or group effects are fixed, the procedures described for CR experiments are applicable for examining and estimating intergroup contrasts with the modification that the ANOVA residuals mean square $s^2 = \hat{V}(\varepsilon_{ij})$ now has $f = (a-1)(b-1)$

instead of $a(n-1)$ degrees of freedom. In particular, the $100(1-\alpha)\%$ confidence interval for the difference between two means, $(\mu_h - \mu_k)$ say, is

$$(\bar{y}_{h.} - \bar{y}_{k.}) \mp t\{(a-1)(b-1); \alpha/2\} \sqrt{\frac{2s^2}{b}} \qquad (107.2)$$

since b observations contribute to each treatment mean.

Example: The data from Table 105, rearranged in treatment by block array, are given in Table 107.2 below.

Computations, in suggested order, are

(i) Correction term $= \dfrac{G^2}{ab}$ $\qquad\qquad = \dfrac{(78)^2}{12} = 507$

(ii) $\displaystyle\sum_i \sum_j y_{ij}^2 \qquad = \qquad 6^2 + 8^2 + \cdots + 6^2 + 4^2 = 544$

(iii) ss(total) $\qquad = \displaystyle\sum\sum y_{ij}^2 - \dfrac{G^2}{ab} \qquad\qquad = 544 - 507 = \quad 37$

(iv) ss(blocks) $\qquad = \dfrac{1}{a}\displaystyle\sum B_j^2 - \left(\dfrac{G^2}{ab}\right) = \dfrac{26^2 + 32^2 + 20^2}{4} - 507 \quad = \quad 18$

(v) ss(treatments) $= \dfrac{1}{b}\displaystyle\sum T_i^2 - \left(\dfrac{G^2}{ab}\right) = \dfrac{18^2 + \cdots + 15^2}{3} - 507 \quad = \quad 15$

(vi) ss(residuals) $\qquad$ = total ss $-$ block ss $-$ treatments ss

$$37 \quad - \quad 18 \quad - \quad 15 \qquad\qquad = \quad 4$$

Hence, with ms = ss/df, the ANOVA in Table 107.3 is obtained.

Table 107.2. Randomized Block Data ($a = 4$, $b = 3$)

Treatment (i)	Block (j)			Total	Mean ($\bar{y}_{i.}$)
	1	2	3		
A	6	8	4	18	6
B	6	9	6	21	7
C	9	9	6	24	8
D	5	6	4	15	5
Total	26	32	20	78(G)	26
Mean ($\bar{y}_{.j}$)	6.5	8	5		6.5($\bar{y}_{..}$)

Table 107.3. ANOVA for Data in Table 107.2

Source	df	ss	ms	F_T
Blocks	2	18	9	
Treatments	3	15	5	7.5[a]
Residuals	6	4	0.67	
Total	11	37		

[a] $0.01 < P < 0.05$.

As shown, the test statistic is $F_T = 5/0.67 = 7.5$ which exceeds $F(3, 6; 0.05) = 4.76$ but is less than $F(3, 6; 0.01) = 9.78$ so that the exceedance probability lies between 0.01 and 0.05. Hence, at $\alpha = 0.05$, real differences between treatment population means are inferred in the fixed treatment effects case, and $\sigma_A^2 > 0$ is inferred if treatment effects are random. As an example of contrast confidence interval estimation, the 95% confidence interval for the difference $(\mu_A - \mu_B)$ between the population means for the first two treatments, is, with $t_{\alpha/2} = t(6; 0.025)$ in (107.2)

$$(\bar{y}_{1.} - \bar{y}_{2.}) \mp t_c \sqrt{\frac{2s^2}{b}} = (6 - 7) \mp 2.447 \sqrt{\frac{2(0.67)}{3}} = -2.6, 0.6 \, \text{lb}$$

§108. THE EFFICIENCY OF RB DESIGNS

As compared with CR designs, RB designs will be more efficient when the reduction in the residual variance achieved by good blocking outweighs the fact that, for the same number of experimental units, the residual variance is estimated with fewer degrees of freedom. The RB ANOVA can be used to estimate s_{ECR}^2, the residual variance of an equivalent CR experiment with the same experimental units, as

$$s_{\text{ECR}}^2 = \frac{(b - 1)(\text{ms blocks}) + b(a - 1)(\text{ms residuals})}{ab - 1} \qquad (108.1)$$

Incorporating a theoretical adjustment for the differing numbers of degrees of freedom, the efficiency of an RB design relative to the equivalent CR design is then calculated as

$$\text{Relative efficiency} = \frac{100(s^2_{ECR}/s^2_{RB})(f_{RB}+1)(f_{CR}+3)}{(f_{RB}+3)(f_{CR}+1)}\% \quad (108.2)$$

with $s^2_{RB} = $ ms residuals $= s^2$ from Table 107.1, $f_{RB} = (a-1)(b-1)$, and $f_{CR} = a(b-1)$.

Example: For the data analyzed in Table 107.3

$$s^2_{ECR} = \frac{(2)(9)+3(3)(0.67)}{11} = 2.18$$

so that

$$\text{Relative efficiency} = \frac{100(2.18/0.67)(7)(11)}{(9)(9)} = 309\%$$

The value here is unusually high; the data are artificial. Otherwise, the result would suggest taking close to $3(309)/100 = 9.3$ replications per treatment in a CR experiment to achieve the precision of the RB experiment with three blocks.

§109. RESIDUALS AND MODEL VERIDICALITY

The model (106.1) shows that the mean of the conceptual population of observations on the ith treatment in the jth block is

$$\mu_{ij} = \mu(y_{ij}) = \mu + \alpha_i + \beta_j$$

of which the estimate is $\hat{\mu} + \hat{\alpha}_i + \hat{\beta}_j$; that is,

$$\hat{y}_{ij} = \bar{y}_{..} + (\bar{y}_{i.} - \bar{y}_{..}) + (\bar{y}_{.j} - \bar{y}_{..}) = \bar{y}_{i.} + \bar{y}_{.j} - \bar{y}_{..}$$

Corresponding to $\varepsilon_{ij} = y_{ij} - \mu_{ij}$, the i,jth residual is then

$$e_{ij} = y_{ij} - \hat{y}_{ij} = (y_{ij} - \bar{y}_{i.} - \bar{y}_{.j} + \bar{y}_{..}) \quad (109.1)$$

Thus, for the data in Table 107.2, the $i=2$, $j=3$ residual is

$$e_{23} = 6-7-5+6.5 = 0.5 \quad \text{and similarly} \quad e_{32} = -0.5$$

The ε_{ij}'s are uncorrelated, while the corresponding e_{ij}'s are not; they sum to zero for each treatment and each block individually and it can be checked that these summations amount to $(a+b-1)$ different constraints on the e_{ij}'s. Hence, the number of degrees of freedom for the residuals sum of squares, which is $\sum_i \sum_j e_{ij}^2$, is the number of e_{ij}'s less the number of constraints, that is, $ab - (a + b - 1) = (a - 1)(b - 1)$.

Residuals are usefully scrutinized for signals concerning the veridicality of the model to see how well it matches the observations and describes the data genesis. For a good match most of the e_{ij}'s should be small. Some possibilities follow.

One extremely large e_{ij} may sometimes be traced to an incorrectly recorded or otherwise fortuitously or even (eureka!) serendipitously discrepant y_{ij} and significance tests are available to examine such e_{ij} outliers (outliars?); see, for example, Barnett and Lewis (1978). Several extremely large residuals may indicate model inadequacy or falseness of the assumptions of independent, normally distributed, constant variance ε_{ij}'s. In dubious cases, a test for normality (e.g., see Martinez and Iglewicz, 1981) may be informative.

The e_{ij} may be plotted against the y_{ij} themselves. Patterns in the plot may again signal model inadequacy. In particular, if the e_{ij} increase with y_{ij}, it may be that $V(\varepsilon_{ij})$ increases with the response so that the regular analysis, assuming constant variance, is invalid. One of the data conversions in §110 may be effective in such cases.

When observations have been collected in a time or space sequence, the residuals may be plotted in the same sequence to see if trends have afflicted the supposedly random residual component. Sets of residuals for each block and treatment separately may also be examined. An extremely large or small set may again put the constant variance assumption in doubt. The simple F_{max}-test for variance heterogeneity is described in Dowdy and Wearden (1983).

Linear model additivity: Models such as (106.1) are linear in the parameters and represent responses as additively composed of effects represented by the parameter terms, such as the overall mean, block and treatment effects, together with an additive residual variability element ε_{ij}. When additivity is in doubt it may be tested by the test for nonadditivity (Tukey, 1949); see also Johnson and Graybill (1972).

In practice ANOVA procedures are often so robust, especially when replication numbers are equal, that moderate departures from protocol model requirements can be ignored without obtaining misleading inferences. But if immoderate departures cannot be regularized, by a data conversion, for example, distribution unspecified procedures are indicated (e.g., see Hollander and Wolfe, 1973).

§110. DATA CONVERSIONS (TRANSFORMATIONS)

For certain types of model invalidity, a transformation may be used to convert each observation into a directly related variate and thus obtain a new data set that better satisfies the basic model assumptions. The regular analysis, comprising the ANOVA, hypothesis tests, and point and interval estimates for means and contrasts, is then carried out on the new data set. Corresponding point and interval estimates, in terms of the original variate units, are obtained by reversing the transformation, as exemplified in §81.

A conversion to reduce variance heterogeneity may also reduce associated non-normality in the data. Some specific variance heterogeneity remedies are noted below. The conversions can be used for data from CR, RB, Latin square, and other investigations.

Variance proportional to the mean: When the variance increases linearly with the mean, the square root conversion is used and, instead of analyzing the original responses y, the linear additive model analysis is carried out on the variate $y^* = \sqrt{y + c}$, where c is a constant. Anscombe (1948), suggested taking $c = 3/8$. This transformation is often appropriate when the basic data are counts, of blood cells or bacteria for example, which follow the Poisson distribution for which it was noted in §44 that the variance is equal to the mean.

Standard deviation proportional to the mean: When the variance increases linearly with the square of the mean, the logarithmic conversion is used and the analysis is carried out on the variate $y^* = \log y$ or, if some y-values are close, or equal, to 0, $y^* = \log(y + 1)$. This conversion is also sometimes used when group or treatment effects on variability are being examined and the analyzed variate is itself a sample variance.

Variance proportional to the fourth power of the mean: Survival time responses sometimes exhibit this type of dependence for which the logarithmic conversion is not strong enough. Preferred is the reciprocal conversion giving the variate analyzed as $y^* = 1/y$ or, if some y-observations are very small or equal to 0, as $y^* = 1/(y + 1)$.

Proportions: If, in the absence of procedures specific for particular experimental contexts, ANOVA procedures are used on proportions (or percentages), data conversion is not commonly made if the proportions lie in the central range from, say, 0.3 to 0.7. If extreme proportions, closer to 0 or 1, are present and the denominators of the ratios are the same or only slightly different, the arcsine transformation is used. For this, the variate analyzed is $y^* = \arcsin\sqrt{y}$, where the y^* corresponding to each proportion y can be read from commonly available statistical tables.

§111. LATIN SQUARE (LS) DESIGNS

Investigation contexts: Randomized block designs achieve precision by removing irrelevant or background variability, that due to block differences, from the residual variability used to examine intertreatment contrasts. Latin square designs extend this principle to situations where two sources of background variability are identifiable. In the simplest form, one observation is obtained on each treatment, at each level or component, of each of the two background variability sources.

Example: Two factors influencing tire wear are (i) the car type and (ii) the wheel position. Wear characteristics of $a = 4$ tire treads could be examined by using four cars of different types so that each tread could be used on each car and at each of the four wheel positions left front, right front, left rear, and right rear.

Randomization: Beginning with a basic LS arrangement with letters denoting treatments, randomization may be effected (see Table A1) by rearranging the rows and columns in random orders and randomly assigning treatments to the letters in the square.

Fixed and random effects: According to the context and the objectives of the investigation, the effects of the levels or components of each of the two background variability sources may be either fixed or random. Treatments may also be fixed or random.

§112. A LATIN SQUARE EXPERIMENT: EXAMPLE

Suppose the following, fictitious, experiment using a LS design was carried out to examine the hypothesis that prolonged intake of *Cannabis sativa* induces premature aging in rats. The treatments were four levels: $A = 0$, $B = 1$, $C = 2$, and $D = 3$ units of *Cannabis* taken daily. Four male rat pups were randomly selected from each of four rat litters to give $4 \times 4 = 16$ experimental subjects. At the end of the feeding regime, each of four observers measured four of the rat pups, one from each litter and

Table 112. A Latin Square Experiment: Design and Data

Observer	Litter Number				Total
	1	2	3	4	
1	A	D	B	C	
	2.8	16.0	10.8	7.8	37.4
2	D	B	C	A	
	3.5	6.0	4.3	1.2	15.0
3	C	A	D	B	
	5.1	4.4	7.3	4.3	21.1
4	B	C	A	D	
	2.1	10.7	7.5	6.8	27.1
Total	13.5	37.1	29.9	20.1	100.6
Treatment	A	B	C	D	Total
Total	15.9	23.2	27.9	33.6	100.6

treatment group, for the aging effect. The design used, wherein the letters show the treatment each rat received, is that in Table 112. The values beneath each letter are the responses to be analyzed.

§113. STATISTICAL MODELS FOR LATIN SQUARES

A typical response $y_{(i)jk}$ is that to the ith treatment in the jth row and the kth column. If treatment, row, and column effects are all fixed, the response is additively modeled as

$$y_{(i)jk} = \mu + \alpha_i + \rho_j + \gamma_k + \varepsilon_{(i)jk} \qquad (113.1)$$

with

$$\sum_i \alpha_i = \sum_j \rho = \sum_k \gamma = 0 \quad \text{and} \quad \varepsilon_{(i)jk} \sim NI(0, \sigma^2)$$

where the subscripts i, j, and k run from $1, \ldots, a$.

Taking $\bar{y}$, $\bar{y}_{Ti}$, $\bar{y}_{Rj}$, and $\bar{y}_{Ck}$ to denote the overall mean, the ith treatment mean, the jth row mean, and the kth column mean, respectively, the parameters, their estimates, and interpretations, are as follows:

Parameter	Estimate	Interpretation
μ	$\bar{y}$	Overall mean
α_i	$\bar{y}_{Ti} - \bar{y}$	Deviation for the ith treatment effect
ρ_j	$\bar{y}_{Rj} - \bar{y}$	Deviation for the jth row component
γ_k	$\bar{y}_{Ck} - \bar{y}$	Deviation for the kth column component

As usual, when any of the treatment, row, and column effects are random, a corresponding roman letter replaces the greek letter in (113.1) and a corresponding variance assumption replaces the restriction that the effects sum to zero. Thus, if treatment effects are random, α_i is replaced by A_i and $\sum \alpha_i = 0$ by $A_i \sim NI(0, \sigma_A^2)$.

Population means of ANOVA mean squares: The population mean of the residual mean square is σ^2. For treatment effects random or fixed the mean of the between-treatments mean square is $\sigma^2 + a\sigma_A^2$ or $\sigma^2 + [a/(a-1)]\sum \alpha_i^2$, respectively. Analogous expressions obtain for the between-rows and between-columns mean squares.

§114. STATISTICAL ANALYSIS FOR LATIN SQUARE EXPERIMENTS

The ANOVA identity is

Total corrected ss = ss(rows) + ss(columns) + ss(treatments) + ss(residual)

so that, as in previous cases, the ANOVA residual ss is obtained by subtraction. The number of degrees of freedom for each of the row, column, and treatment ss is $(a-1)$ so that the residual ss has

$$a^2 - 1 - 3(a-1) = (a-1)(a-2) \text{ df}$$

With T_i, R_j, and C_k as the ith treatment, jth row, and kth column totals and G as the grand total, the ANOVA is computed as in Table 114.1.

Hypothesis tests: For fixed treatment effects, $H_T: \alpha_1 = \cdots = \alpha_a = 0$, H_A: not H_T and, for random treatment effects, $H_T: \sigma_A^2 = 0$, $H_A: \sigma_A^2 > 0$

Table 114.1. ANOVA for a Latin Square Experiment on a Treatments

Variability Source	df	ss	ms
Between-rows	$a-1$	$\dfrac{1}{a}\sum R_j^2 - \left(\dfrac{G}{a}\right)^2$	s_3^2
Between-columns	$a-1$	$\dfrac{1}{a}\sum C_k^2 - \left(\dfrac{G}{a}\right)^2$	s_2^2
Between-treatments	$a-1$	$\dfrac{1}{a}\sum T_i^2 - \left(\dfrac{G}{a}\right)^2$	s_1^2
Residuals	$(a-1)(a-2)$	By subtraction	s^2
Total	a^2-1	$\sum_j \sum_k y_{(i)jk}^2 - \left(\dfrac{G}{a}\right)^2$	

are both examined by comparing the test statistic

$$F_T = \frac{\text{ms between treatments}}{\text{ms residual}} = \frac{s_1^2}{s^2}$$

with critical regions from the $F\{(a-1),(a-1)(a-2)\}$-distribution.

Further analysis: As with RB designs, further analysis for the treatments random situation depends on the experimental context and objectives. When treatment effects are fixed, appropriate intertreatment contrast procedures (§94ff) are applicable. In particular, for example, the $100(1-\alpha)\%$ confidence interval for the difference between the mean parameters for treatment 1 and treatment 2 is calculated as

$$(\bar{y}_1 - \bar{y}_2) \mp t_{\alpha/2} \sqrt{\frac{2s^2}{a}}, \qquad t_{\alpha/2} = t\{(a-1)(a-2);\ \alpha/2\}$$

Example: The ANOVA for the data in Table 112, calculated in accordance with Table 114.1, is given in Table 114.2.

As a natural planned contrast in this context [see Example (ii), §94], the mean linear effect of response on ingestion level can be estimated and tested as in §95 and §96. The estimate of the contrast, that is, of the regression coefficient of *mean* response ($\bar{y}$) on level (x), is

$$l = \frac{\sum (x-\bar{x})\bar{y}}{\sum (x-\bar{x})^2}$$

Table 114.2. ANOVA for LS Experiment in Table 112

Source	df	ss	ms	F_T
Rows	3	68.32	22.77	
Columns	3	81.65	27.22	
Treatments	3	42.09	14.03	3.6^a
Residuals	6	23.46	3.91	
Total	15	215.52		

$^a 0.10 > P > 0.05.$

For $x = 0, 1, 2,$ and 3 for treatments A, B, C, and D, the values of $(x - \bar{x})$ are $-3/2, -1/2, 1/2,$ and $3/2$, so that $\sum (x - \bar{x}) = 0$ as required, and

$$l = \frac{-(3/2)\bar{y}_A - (1/2)\bar{y}_B + (1/2)\bar{y}_C + (3/2)\bar{y}_D}{2\{(9/4) + (1/4)\}}$$

$$= \frac{-3T_A - T_B + T_C + 3T_D}{(8)(2)(10/4)}$$

$$= \frac{-3(15.9) - 23.2 + 27.9 + 3(33.6)}{40}$$

$$= 1.44 \text{ aging units/ingestion unit.}$$

With $L = -3T_A - T_B + T_C + 3T_D = 57.8$, the test statistic for

$$H_T: \mu(l) = \lambda = 0, \quad H_A: \lambda \neq 0$$

is

$$t_T = \frac{L}{\sqrt{\hat{V}(L)}} = \frac{57.8}{\sqrt{4(20)s^2}} = \frac{57.8}{\sqrt{80(3.91)}} = 3.27$$

which exceeds $t_{\alpha/2} = t(6; 0.025) = 2.447$ and $\alpha = 0.05$ indicates a non-zero regression coefficient. Finally, since $l = L/40$, the 95% confidence interval for the regression coefficient is

$$\frac{\{57.8 \mp 2.447\sqrt{80(3.91)}\}}{40} = 0.4, 2.5 \text{ aging units/ingestion unit}$$

§115. RESIDUALS IN LS EXPERIMENTS

The remarks on RB residuals in §109 also apply to LS residuals. For these, replacing parameters on the right-hand side of

$$\varepsilon_{(i)jk} = y_{(i)jk} - \mu - \alpha_i - \rho_i - \gamma_k$$

by their estimates (§113) gives the residual

$$e_{(i)jk} = y_{(i)jk} - \bar{y}_{Ti} - \bar{y}_{Rj} - \bar{y}_{Ck} + 2\bar{y}$$

As an example, the residual for the observer 2, litter 3 unit which received treatment C is, from Table 112,

$$e_{(3)23} = 4.3 - \frac{27.9}{4} - \frac{15.0}{4} - \frac{29.9}{4} + 2\left(\frac{100.6}{16}\right) = -1.3$$

§116. THE EFFICIENCY OF LS DESIGNS

F-tests, comparing the mean squares for rows and for columns with the residual mean square, may be made to check whether or not inclusion of the corresponding parameters appreciably reduces the residual variance. The efficiency of a LS design relative to a comparable RB design in which one of the row and column classifications is used to provide blocks, the other being ignored, can be estimated. Using rows, for example, as blocks, the residual mean square of the comparable RB analysis has been theoretically estimated to be s_{ERB}^2, where

$$s_{\text{ERB}}^2 = \frac{(\text{ms between columns}) + (a-1)(\text{ms residuals})}{a} \quad (116.1)$$

The corresponding number of degrees of freedom for the residuals variance will be

$$f_{\text{ERB}} = f_{\text{LS}} + (a-1) = (a-1)(a-2) + (a-1) = (a-1)^2$$

The percentage relative efficiency of the LS design is then the ratio $100 s_{\text{ERB}}^2 / s_{\text{LS}}^2$ multiplied by the degrees of freedom adjustment factor $(f_{\text{LS}} + 1)(f_{\text{ERB}} + 3)/(f_{\text{LS}} + 3)(f_{\text{ERB}} + 1)$.

Example: From Table 114.2,

$$s_{\text{ERB}}^2 = \frac{27.22 + 3(391)}{4} = 9.74$$

and the degrees of freedom factor, for $a = 4$, is

$$\frac{\{(a-1)(a-2)+1\}\{(a-1)^2+3\}}{\{(a-1)(a-2)+3\}\{(a-1)^2+1\}} = \frac{84}{90}$$

Hence, compared to a RB design using the row classification as blocks, the relative efficiency of the LS design is

$$\text{Relative efficiency} = \frac{100(9.74)(84)}{(3.91)(90)} = 232.5\%$$

With the between-rows mean square instead of the between-columns mean square in (116.1) it may similarly be calculated that, relative to a RB design with columns as blocks, the efficiency estimate of the LS design is $100(8.625)(84)/(3.91)(90) = 205.9\%$. The latter result, for example, suggests that to achieve the same sensitivity as the LS design, a comparable RB experiment with columns as blocks would need at least $2.06 \times 16 = 33$ instead of the 16 experimental units.

§117. MISSING OBSERVATIONS

When at least one of the complete data set for a RB or LS investigation is missing, the consequent imbalance invalidates the regular analyses described above. Procedures using formulas to estimate the missing values are described in statistical methods texts. Texts may also be consulted for an alternative and more convenient procedure that uses the analysis of covariance. For either approach, the assumption is required that the missing values are unrelated to the effects of the treatments under study.

Introduction to Matrix Operations

§118. CONTEXT AND DEFINITIONS

Multiple regression involves statistical procedures for relating a dependent variate y to not one (as in §46ff) but several predictor variables. The principles and computations for multiple regression analyses are considerably simplified by the use of matrices. Matrix methods are also useful in the analysis of data from RB, LS, and other experimental designs (Searle, 1982).

Definitions: A real $r \times c$ matrix, $\mathbf{X}$ say, consists of rc numbers, called *elements* or *scalars*, rectangularly arranged in r rows and c columns. A *square* matrix has $r = c$.

If the element in the ith row and jth column is the number x_{ij}, examples of a 2×3 matrix are

$$\mathbf{X}_1 = \begin{bmatrix} 1 & 4 & 7 \\ 1 & -3 & 2 \end{bmatrix}, \qquad \mathbf{X}_2 = \begin{bmatrix} x_{11} & x_{12} & x_{13} \\ x_{21} & x_{22} & x_{23} \end{bmatrix}$$

so that, in $\mathbf{X}_1$, $x_{11} = 1$, $x_{12} = 4, \ldots, x_{23} = 2$.

Row vector: The simple matrix consisting of $r = 1$ row and c columns is called a row vector. Examples are,

$$\mathbf{X}_3 = [2.1 \quad 3.6 \quad 0.4 \quad 3.2], \qquad \mathbf{X}_4 = [x_1 \quad x_2 \quad \ldots \quad x_c]$$

Column vector: The simple matrix having r rows and just $c = 1$ column is a column vector. Examples are

$$\mathbf{Y}_1 = \begin{bmatrix} 1 \\ 3 \\ 5 \\ 7 \end{bmatrix}, \qquad \mathbf{Y}_2 = \begin{bmatrix} y_1 \\ y_2 \\ \vdots \\ y_r \end{bmatrix}$$

§119. TRANSPOSITION AND SYMMETRY

The *transpose* of an $r \times c$ matrix, $\mathbf{X}$ say, is a $c \times r$ matrix, denoted by $\mathbf{X}'$, obtained by rewriting the elements in the *i*th *row* of $\mathbf{X}$ to be those in the *i*th *column* of $\mathbf{X}'$, $i = 1, \ldots, r$. Examples using the matrices defined above are

$$\mathbf{X}_1' = \begin{bmatrix} 1 & 4 & 7 \\ 1 & -3 & 2 \end{bmatrix}' = \begin{bmatrix} 1 & 1 \\ 4 & -3 \\ 7 & 2 \end{bmatrix}, \quad \mathbf{X}_2' = \begin{bmatrix} x_{11} & x_{12} & x_{13} \\ x_{21} & x_{22} & x_{23} \end{bmatrix}' = \begin{bmatrix} x_{11} & x_{21} \\ x_{12} & x_{22} \\ x_{13} & x_{23} \end{bmatrix}$$

$$\mathbf{X}_4' = \begin{bmatrix} x_1 \\ x_2 \\ \vdots \\ x_c \end{bmatrix}, \qquad\qquad \mathbf{Y}_2' = \begin{bmatrix} y_1 & y_2 & \cdots & y_r \end{bmatrix}$$

and, of course, for any matrix $\mathbf{M}$, $[\mathbf{M}']' = \mathbf{M}$.

Symmetric matrices: A matrix in which the element x_{ij} in the *i*th row and *j*th column is the same as that x_{ji} in the *j*th row and the *i*th column, for all i and j, is termed *symmetric*.

If $\mathbf{X}$ is symmetric, it is unchanged by transposition, so that $\mathbf{X} = \mathbf{X}'$. Symmetric matrices are accordingly necessarily square. Examples are

$$\mathbf{X}_5 = \begin{bmatrix} 2 & 7 & 6 & 4 \\ 7 & 3 & 0 & 9 \\ 6 & 0 & 5 & 1 \\ 4 & 9 & 1 & 8 \end{bmatrix}, \quad \mathbf{M} = \begin{bmatrix} a_1 & a_2 & a_3 \\ a_2 & b_2 & b_3 \\ a_3 & b_3 & c_3 \end{bmatrix}$$

§120. MATRIX ADDITION AND SUBTRACTION

Example:

$$\begin{bmatrix} 1 & 4 & 7 \\ 1 & -3 & 2 \end{bmatrix} + \begin{bmatrix} 1 & -3 & 4 \\ 2 & 0 & -6 \end{bmatrix} - \begin{bmatrix} 2 & -4 & 7 \\ -3 & -3 & 2 \end{bmatrix}$$

$$= \begin{bmatrix} 1+1-2 & 4-3-(-4) & 7+4-7 \\ 1+2-(-3) & -3+0-(-3) & 2-6-2 \end{bmatrix} = \begin{bmatrix} 0 & 5 & 4 \\ 6 & 0 & -6 \end{bmatrix}$$

The example shows the rule that the i,jth element of the result matrix is obtained by summing, or differencing for subtraction, the i,jth elements of the individual matrices. Clearly, all the matrices must have the same number r of rows and c of columns. Examples are:

(i) If

$$\mathbf{X} = \begin{bmatrix} x_{11} & x_{12} \\ x_{21} & x_{22} \\ x_{31} & x_{32} \end{bmatrix} \quad \text{and} \quad \mathbf{Y} = \begin{bmatrix} y_{11} & y_{12} \\ y_{21} & y_{22} \\ y_{31} & y_{32} \end{bmatrix}$$

then

$$\mathbf{X} + \mathbf{Y} = \begin{bmatrix} (x_{11} + y_{11}) & (x_{12} + y_{12}) \\ (x_{21} + y_{21}) & (x_{22} + y_{22}) \\ (x_{31} + y_{31}) & (x_{32} + y_{32}) \end{bmatrix}$$

from which it is also appreciated that for any two matrices, $\mathbf{A}$ and $\mathbf{B}$ say,

$$[\mathbf{A} + \mathbf{B}]' = \mathbf{A}' + \mathbf{B}' \quad \text{and} \quad [\mathbf{A} - \mathbf{B}]' = \mathbf{A}' - \mathbf{B}' \tag{120.1}$$

so that transposed matrices are, naturally, added and subtracted according to the above rule.

(ii) The transpose of the difference between the column vectors,

$$\mathbf{Y} = [y_1 \quad y_2 \quad \cdots \quad y_n]' \quad \text{and} \quad \bar{\mathbf{Y}} = [\bar{y} \quad \bar{y} \quad \cdots \quad \bar{y}]'$$

is the row vector

$$[\mathbf{Y} - \bar{\mathbf{Y}}]' = [(y_1 - \bar{y}) \quad (y_2 - \bar{y}) \quad \cdots \quad (y_n - \bar{y})] \tag{120.2}$$

§121. MATRIX MULTIPLICATIONS

Scalar multiplication: When a matrix is multiplied by a scalar, each of the matrix elements is multiplied by the scalar. For example,

$$(7) \begin{bmatrix} 1 & 3 & 4 \\ 2 & 0 & 6 \end{bmatrix} = \begin{bmatrix} 7 & 21 & 28 \\ 14 & 0 & 42 \end{bmatrix}, \quad (c) \begin{bmatrix} x_{11} & x_{12} \\ x_{21} & x_{22} \end{bmatrix} = \begin{bmatrix} cx_{11} & cx_{12} \\ cx_{21} & cx_{22} \end{bmatrix}$$

Row vector × column vector: The product of a row vector and a column vector is a *scalar* obtained as the sum of the products of pairs of corresponding elements. Examples are:

(i)

$$[1 \quad -3 \quad 4 \quad 7] \begin{bmatrix} 2 \\ 0 \\ 3 \\ -1 \end{bmatrix} = 1(2) - 3(0) + 4(3) + 7(-1) = 7$$

(ii) If $\mathbf{Y} = [y_1 \quad y_2 \quad \cdots \quad y_n]'$, then $\mathbf{Y}' = [y_1 \quad y_2 \quad \cdots \quad y_n]$ and

$$\mathbf{Y}'\mathbf{Y} = [y_1 \quad y_2 \quad \cdots \quad y_n]\begin{bmatrix} y_1 \\ y_2 \\ \vdots \\ y_n \end{bmatrix} = y_1^2 + y_2^2 + \cdots + y_n^2 = \sum_1^n y_i^2$$

(iii) With $[\mathbf{X} - \bar{\mathbf{X}}]$ and $[\mathbf{Y} - \bar{\mathbf{Y}}]$ defined as in (120.2)

$$[\mathbf{X} - \bar{\mathbf{X}}]'[\mathbf{Y} - \bar{\mathbf{Y}}] = (x_1 - \bar{x})(y_1 - \bar{y}) + \cdots + (x_n - \bar{x})(y_n - \bar{y}) = \sum{}' xy$$

Furthermore, using (120.1), $[\mathbf{X} - \bar{\mathbf{X}}]' = \mathbf{X}' - \bar{\mathbf{X}}'$ so that

$$\sum{}' xy = [\mathbf{X} - \bar{\mathbf{X}}]'[\mathbf{Y} - \bar{\mathbf{Y}}] = \mathbf{X}'[\mathbf{Y} - \bar{\mathbf{Y}}] - \bar{\mathbf{X}}'[\mathbf{Y} - \bar{\mathbf{Y}}]$$

$$= \mathbf{X}'\mathbf{Y} - \mathbf{X}'\bar{\mathbf{Y}} - \sum \bar{x}(y_i - \bar{y})$$

$$= \sum x_i y_i - \sum x_i \bar{y} - 0$$

$$= \sum x_i y_i - n\bar{x}\bar{y}$$

This proof of (49.1), in matrix notation, follows that for (2.7).

General matrix multiplication: Taking an example first,

$$[\mathbf{A}][\mathbf{B}] = \begin{bmatrix} 4 & 2 \\ 3 & -1 \end{bmatrix} * \begin{bmatrix} 3 & 5 & 6 \\ 0 & 7 & 2 \end{bmatrix} = \begin{bmatrix} 4(3)+2(0) & 4(5)+2(7) & 4(6)+2(2) \\ 3(3)-1(0) & 3(5)-1(7) & 3(6)-1(2) \end{bmatrix}$$

$$= \begin{bmatrix} 12 & 34 & 28 \\ 9 & 8 & 16 \end{bmatrix} = [\mathbf{AB}]$$

In general, if $\mathbf{M}_1$ and $\mathbf{M}_2$ are any two matrices, they can only be multiplied to obtain the product $\mathbf{M}_1\mathbf{M}_2$ if they are "conformable," that is, if the number of *columns* of $\mathbf{M}_1$ is the same as the number of *rows* of $\mathbf{M}_2$. Then, if $\mathbf{M}_1$ has r_1 rows and c_1 columns and $\mathbf{M}_2$ has c_1 rows and c_2 columns, $\mathbf{M}_1\mathbf{M}_2$ has the same number of rows r_1, as $\mathbf{M}_1$, and the same number of columns, c_2, as $\mathbf{M}_2$. And the i,jth element of $\mathbf{M}_1\mathbf{M}_2$ is the product, a scalar, of the ith row of $\mathbf{M}_1$ with the jth column of $\mathbf{M}_2$ for $i = 1, \ldots, r_1$ and $j = 1, \ldots, c_2$. Thus, in the above example, the 2,3rd element of $[\mathbf{AB}]$ is obtained as the product of the second row of $\mathbf{A}$ and the third column of $\mathbf{B}$. Thus,

$$[3 \quad -1]\begin{bmatrix} 6 \\ 2 \end{bmatrix} = 3(6) - 1(2) = 16$$

In matrix multiplication the order in which the matrices being multiplied occur is important. Except for special circumstances, the general fact is that $[\mathbf{M}_1\mathbf{M}_2] \neq [\mathbf{M}_2\mathbf{M}_1]$. Thus,

$$\begin{bmatrix} 4 & 2 \\ 3 & -1 \end{bmatrix}\begin{bmatrix} 3 & 5 \\ 0 & 7 \end{bmatrix} = \begin{bmatrix} 12 & 34 \\ 9 & 8 \end{bmatrix} \neq \begin{bmatrix} 3 & 5 \\ 0 & 7 \end{bmatrix}\begin{bmatrix} 4 & 2 \\ 3 & -1 \end{bmatrix} = \begin{bmatrix} 27 & 1 \\ 21 & -7 \end{bmatrix}$$

Accordingly, "premultiplying" $\mathbf{M}_2$ by $\mathbf{M}_1$ gives $\mathbf{M}_1\mathbf{M}_2$ and "postmultiplying" $\mathbf{M}_2$ by $\mathbf{M}_1$ gives $\mathbf{M}_2\mathbf{M}_1$.

Column × row multiplication: Another particular case of the preceding general rule occurs when an $r \times 1$ column vector is multiplied by a $1 \times c$ row vector to obtain an $r \times c$ matrix. Examples are:

$$\begin{bmatrix} 3 \\ 0 \\ 7 \end{bmatrix}\begin{bmatrix} 2 & 0 & 5 & 4 \end{bmatrix} = \begin{bmatrix} 6 & 0 & 15 & 12 \\ 0 & 0 & 0 & 0 \\ 14 & 0 & 35 & 28 \end{bmatrix}$$

$$\boldsymbol{\varepsilon}\boldsymbol{\varepsilon}' = \begin{bmatrix} \varepsilon_1 \\ \varepsilon_2 \\ \cdot \\ \cdot \\ \cdot \\ \varepsilon_n \end{bmatrix}\begin{bmatrix} \varepsilon_1 & \varepsilon_2 & \cdots & \varepsilon_n \end{bmatrix} = \begin{bmatrix} \varepsilon_1^2 & \varepsilon_1\varepsilon_2 & \cdots & \varepsilon_1\varepsilon_n \\ \varepsilon_2\varepsilon_1 & \varepsilon_2^2 & \cdots & \varepsilon_2\varepsilon_n \\ & & \cdots & \\ \varepsilon_n\varepsilon_1 & \varepsilon_n\varepsilon_2 & \cdots & \varepsilon_n^2 \end{bmatrix}$$

§122. TRANSPOSES OF MATRIX PRODUCTS

The transpose of a matrix product is equal to the product of the individual matrices, each transposed, written in the reverse order. Thus, from the $[\mathbf{AB}]$ product example above,

$$[\mathbf{AB}]' = \begin{bmatrix} 12 & 34 & 28 \\ 9 & 8 & 16 \end{bmatrix}' = \begin{bmatrix} 12 & 9 \\ 34 & 8 \\ 28 & 16 \end{bmatrix} = \begin{bmatrix} 3 & 0 \\ 5 & 7 \\ 6 & 2 \end{bmatrix}\begin{bmatrix} 4 & 3 \\ 2 & -1 \end{bmatrix} = \mathbf{B}'\mathbf{A}'$$

The rule extends to products of more than two matrices, so that,

$$[\mathbf{M}_1\mathbf{M}_2 \cdots \mathbf{M}_k]' = [\mathbf{M}_k' \cdots \mathbf{M}_2'\mathbf{M}_1'] \qquad (122.1)$$

The rule shows that some matrix products are necessarily symmetric matrices. Thus,

$$[\mathbf{X}'\mathbf{X}]' = [\mathbf{X}'][\mathbf{X}']' = [\mathbf{X}'\mathbf{X}]$$

and, if $\mathbf{M}$ is any symmetric matrix, $[\mathbf{X}'\mathbf{M}\mathbf{X}]$ is symmetric because

$$[\mathbf{X}'\mathbf{M}\mathbf{X}]' = [\mathbf{X}'\mathbf{M}'\mathbf{X}] = [\mathbf{X}'\mathbf{M}\mathbf{X}]$$

Row × column products, being scalars, are necessarily symmetric. Thus,

$$[4 \quad 1 \quad 3][2 \quad 7 \quad 2]' = 21 = [2 \quad 7 \quad 2][4 \quad 1 \quad 3]'$$

§123. UNIT OR IDENTITY MATRICES

The square matrix in which each element in the principal diagonal is unity and all other elements are zero is $\mathbf{I}_r$, the $r \times r$ unit matrix. The successive unit matrices are

$$\mathbf{I}_1 = 1, \qquad \mathbf{I}_2 = \begin{bmatrix} 1 & 0 \\ 0 & 1 \end{bmatrix}, \qquad \mathbf{I}_3 = \begin{bmatrix} 1 & 0 & 0 \\ 0 & 1 & 0 \\ 0 & 0 & 1 \end{bmatrix} \cdots$$

generally written as just $\mathbf{I}$ when the dimensions are clear.

In matrix algebra, $\mathbf{I}$ plays the same role as unity does in arithmetic. In particular, any matrix pre- or postmultiplied by $\mathbf{I}$ remains unchanged. Thus,

$$[\mathbf{I}]\begin{bmatrix} 4 & 2 \\ 3 & 1 \end{bmatrix} = \begin{bmatrix} 1 & 0 \\ 0 & 1 \end{bmatrix}\begin{bmatrix} 4 & 2 \\ 3 & 1 \end{bmatrix} = \begin{bmatrix} 4 & 2 \\ 3 & 1 \end{bmatrix}$$

and

$$\begin{bmatrix} x_{11} & x_{12} \\ x_{21} & x_{22} \end{bmatrix}[\mathbf{I}] = \begin{bmatrix} x_{11} & x_{12} \\ x_{21} & x_{22} \end{bmatrix}\begin{bmatrix} 1 & 0 \\ 0 & 1 \end{bmatrix} = \begin{bmatrix} x_{11} & x_{12} \\ x_{21} & x_{22} \end{bmatrix}$$

and, generally, for any $r \times c$ matrix $\mathbf{X}$,

$$[\mathbf{I}_r][\mathbf{X}] = [\mathbf{X}][\mathbf{I}_c] = [\mathbf{X}]$$

§124. THE INVERSE OF A SQUARE MATRIX

If two square matrices $\mathbf{A}$ and $\mathbf{B}$ are such that $\mathbf{AB} = \mathbf{I}$, $\mathbf{B}$, written as $\mathbf{A}^{-1}$, is the inverse of $\mathbf{A}$ and $\mathbf{A}$, written as $\mathbf{B}^{-1}$, is the inverse of $\mathbf{B}$ so that, generally,

$$\mathbf{AA}^{-1} = \mathbf{A}^{-1}\mathbf{A} = \mathbf{I}$$

Example: Because

$$\begin{bmatrix} 4 & 2 \\ 3 & 2 \end{bmatrix}\begin{bmatrix} 1 & -1 \\ -1.5 & 2 \end{bmatrix} = \begin{bmatrix} 1 & 0 \\ 0 & 1 \end{bmatrix}$$

$$\begin{bmatrix} 4 & 2 \\ 3 & 2 \end{bmatrix}^{-1} = \begin{bmatrix} 1 & -1 \\ -1.5 & 2 \end{bmatrix} \quad \text{and} \quad \begin{bmatrix} 1 & -1 \\ -1.5 & 2 \end{bmatrix}^{-1} = \begin{bmatrix} 4 & 2 \\ 3 & 2 \end{bmatrix}$$

Singular matrix: A singular matrix is one that has no inverse. An example is the matrix

$$\begin{bmatrix} 4 & 2 \\ 2 & 1 \end{bmatrix}$$

for which no multiplier matrix with finite elements can be found to give a unit matrix for the product.

Except for some special matrices, matrix inverses are generally obtained using computer routines.

§125. THE INVERSES OF (i) 2×2 MATRICES, (ii) DIAGONAL MATRICES, (iii) BORDERED MATRICES, AND (iv) MATRIX PRODUCTS

(i) If

$$\mathbf{M} = \begin{bmatrix} a & b \\ c & d \end{bmatrix}$$

then

$$\mathbf{M}^{-1} = \frac{1}{(ad - bc)} \begin{bmatrix} d & -b \\ -c & a \end{bmatrix}$$

unless $ad = bc$ in which case $\mathbf{M}$ is singular.

(ii) A **diagonal matrix** $\mathbf{D}$ is one in which all the elements, except those in the principal diagonal, are zero. If the successive elements in the principal diagonal are $d_1, d_2, \ldots, d_p$, then $\mathbf{D}^{-1}$ is also diagonal with successive diagonal elements $1/d_1, 1/d_2, \ldots, 1/d_p$, provided none of the d_i's is zero in which case $\mathbf{D}$ is singular. Thus, with $d_1 = 4$, $d_2 = 1$, and $d_3 = 5$,

$$\mathbf{D} = \begin{bmatrix} 4 & 0 & 0 \\ 0 & 1 & 0 \\ 0 & 0 & 5 \end{bmatrix} \quad \text{and} \quad \mathbf{D}^{-1} = \begin{bmatrix} 1/4 & 0 & 0 \\ 0 & 1 & 0 \\ 0 & 0 & 1/5 \end{bmatrix}$$

(iii) A **bordered matrix** is here taken to have a nonzero element in the 1,1th position and zeros for all other elements in the first row and the first column. The inverse of a bordered square matrix $\mathbf{M}$ is also a bordered matrix and consists of the inverse of the matrix obtained by deleting the first row and column of $\mathbf{M}$, bordered by a first row and first column of zero elements except again for the 1,1th element which is the reciprocal of the 1,1th element of **M**. Examples are:

(a) If

$$\mathbf{M} = \begin{bmatrix} 5 & 0 & 0 \\ 0 & 4 & 2 \\ 0 & 3 & 2 \end{bmatrix}, \quad \text{then} \quad \mathbf{M}^{-1} = \begin{bmatrix} 1/5 & 0 & 0 \\ 0 & 1 & -1 \\ 0 & -1.5 & 2 \end{bmatrix}$$

wherein

$$\begin{bmatrix} 1 & -1 \\ -1.5 & 2 \end{bmatrix} = \begin{bmatrix} 4 & 2 \\ 3 & 2 \end{bmatrix}^{-1}$$

(b) If **M** is the $r \times r$ matrix below and the submatrix **A** is nonsingular

$$\mathbf{M}_{(r \times r)} = \begin{bmatrix} n & 0 & \cdots & 0 \\ 0 & & & \\ \vdots & & \mathbf{A} & \\ \vdots & & (r-1) \times (r-1) & \\ 0 & & & \end{bmatrix}, \quad \text{and} \quad \mathbf{M}^{-1}_{(r \times r)} = \begin{bmatrix} 1/n & 0 & \cdots & 0 \\ 0 & & & \\ \vdots & & \mathbf{A}^{-1} & \\ \vdots & & (r-1) \times (r-1) & \\ 0 & & & \end{bmatrix}$$

(iv) **Product matrices:** If **M** is the product of nonsingular matrices, $\mathbf{M}^{-1}$ is the product of their inverses written in reverse order. Thus, if

$$\mathbf{M} = \mathbf{AB} \cdots \mathbf{JK}, \quad \text{then} \quad \mathbf{M}^{-1} = \mathbf{K}^{-1}\mathbf{J}^{-1} \cdots \mathbf{B}^{-1}\mathbf{A}^{-1}$$

Taking a product of just two matrices exemplifies why. For, if $\mathbf{M} = \mathbf{AB}$,

$$\mathbf{MM}^{-1} = \mathbf{ABB}^{-1}\mathbf{A}^{-1} = \mathbf{AIA}^{-1} = \mathbf{AA}^{-1} = \mathbf{I}$$

Multiple Regression: Describing Data in Terms of Several Variables (Plus Chance)

§126. INVESTIGATION CONTEXT

Multiple regression procedures extend those in §46ff, for one, fixed, predictor variable, to cases where the response variate y depends on $p \geqq 1$ predictor variables $x_1, x_2, \ldots, x_p$. For example, observations might be taken to investigate the prediction of y, the distance required to stop a car, on predictor variables such as $x_1 = (\text{car speed})^2$, $x_2 = \text{car weight}$, $x_3 = \text{tire wear}$, $x_4 = \text{road surface coefficient of friction}$, $x_5 = \text{brake lining contact area}$, and $x_6 = \text{driver reflex time}$. For the strict applicability of the procedures to be described it is to be noted that, as in §47, regression Model 1, the predictor variable values should either be entirely free of measurement error or so "controlled" that measurement error is effectively negligible.

§127. THE DATA AND THE STATISTICAL MODEL

In the simplest case it is convenient to regard a single "observation" as a collection of $(p + 1)$ interdependent measurements, one on the response variate and one on each of the predictor variables, all measured jointly on one occasion or in one "experiment." The complete data set consists of n such mutually independent "observations," or data points, which can be written

$$(y_i, x_{1i}, x_{2i}, \ldots, x_{pi}), \qquad i = 1, \ldots, n \qquad (127.1)$$

Thus, in §128, Example (ii), where $p = 2$ and $n = 7$, the third observation is ($y_3 = 5.5$, $x_{13} = 5.0$, $x_{23} = 6.0$) and the three values can be regarded as the coordinates of a point in three dimensions referred to axes for y, x_1, and x_2.

Considering the ith observation (127.1), if the values of all the predictor variables could be held fixed at $x_{1i}, x_{2i}, \ldots, x_{pi}$ and repeated response values $y_{i1}, y_{i2}, \ldots$ were observed, the aggregate of all such values is the conceptual population of responses at these fixed x-values. The one y_i response actually observed is regarded as a random member of this population. The mean of this population is

$$\mu(y, \text{ given that } x_1 = x_{1i} \text{ and } x_2 = x_{2i} \cdots \text{ and } x_p = x_{pi})$$

Using the symbol "$|$" to denote "given that" or "for" or "when" this mean is notationally abbreviated to

$$\mu(y \mid x_{1i}, x_{2i}, \ldots, x_{pi}) = \mu(y \mid \mathbf{X}_i)$$

and to this stage the observations are modeled as

$$y_i = \mu(y \mid \mathbf{X}_i) + \varepsilon_i, \qquad V(\varepsilon_i) = \sigma^2, \qquad i = 1, \ldots, n \qquad (127.2)$$

wherein $V(\varepsilon_i) = V(y \mid \mathbf{X}_i)$ is the variance of each of the above conceptual populations of y's at the n respective sets of $\mathbf{X}_i = (x_{1i} \cdots x_{pi})$-values. Except that the subscript has now been omitted for brevity, this σ^2 corresponds to that in (47.3), $\sigma^2_{y|x}$, and again, as in the $p = 1$ case there, we assume that this variance is the same for all the n data point situations.

The structural part of the model relates the population means $\mu(y \mid \mathbf{X})$ to values of the predictor variables as

$$\mu(y \mid \mathbf{X}) = \beta_0 + \beta_1 x_1 + \cdots + \beta_p x_p \qquad (127.3)$$

so that, for the observed points,

$$\mu(y \mid \mathbf{X}_i) = \beta_0 + \beta_1 x_{1i} + \cdots + \beta_p x_{pi}, \qquad i = 1, \ldots, n \qquad (127.4)$$

From (127.3) it is seen that β_0 is the mean, $\mu(y \mid x_1 = \cdots = x_p = 0)$, of responses conceptually observable when all the predictor variables are zero; β_0 is therefore the intercept parameter corresponding to that in (46.3), α, for the $p = 1$ case.

The partial regression coefficients of y on x_j are the parameters β_j, $j = 1, \ldots, p$ in (127.3) and β_1, the partial regression coefficient of y on x_1, for example, measures the amount by which the population mean of the response variate changes for a one-unit change in the predictor variable x_1 when all the other predictor variables are, conceptually, held constant. Thus, if the variables $x_2, \ldots, x_p$ are held to the constant values $c_2, \ldots, c_p$, respectively, (127.3) becomes

$$\mu(y \mid \mathbf{X}) = \beta_0 + \beta_1 x_1 + \beta_2 c_2 + \cdots + \beta_p c_p$$

$$= \left(\beta_0 + \sum_2^p \beta_j c_j\right) + \beta_1 x_1$$

as in the simple linear regression situation (46.1) with the intercept

$$\alpha = \left(\beta_0 + \sum_2^p \beta_j c_j\right)$$

Note: It is important to distinguish the *partial* regression coefficient of y on x_j, which differs both in value and interpretation, from the *simple* linear regression coefficient obtained by *ignoring* all but the x_j predictor variable, that is, by entirely omitting all the predictor variables except x_j from the model. The distinction is sometimes emphasized notationally: if $p = 4$, for example, $\beta_{y2|134}$ can be written for the partial regression coefficient of y on x_2.

The statistical model: Bringing (127.2) and (127.4) together and incorporating the normality and independence assumptions for the ε_i, required for hypothesis testing and estimation procedures, we obtain the multiple regression statistical model for the observations as

$$y_i = \beta_0 + \beta_1 x_{1i} + \cdots + \beta_p x_{pi} + \varepsilon_i \qquad (127.5)$$

$$\varepsilon_i \sim NI(0, \sigma^2), \qquad i = 1, \ldots, n$$

Although the model (127.5) is still widely used, computational advantages are demonstrable for the alternative, equivalent, model

$$y_i = \mu + \beta_1 (x_{1i} - \bar{x}_1) + \cdots + \beta_p (x_{pi} - \bar{x}_p) + \varepsilon_i \qquad (127.6)$$

$$\varepsilon_i \sim NI(0, \sigma^2), \qquad i = 1, \ldots, n$$

wherein the value of each predictor variable is first "centered" by subtracting its mean; thus, instead of x_{ji} in (127.5), we have $(x_{ji} - \bar{x}_j)$, $\bar{x}_j = \sum_1^n x_{ji}/n$, in (127.6), for $j = 1, \ldots, p$. The partial regression coefficients are exactly the same and have the same interpretation in both models which are connected by noting that

$$\mu = \beta_0 + \beta_1 \bar{x}_1 + \cdots + \beta_p \bar{x}_p \qquad (127.7)$$

where μ is simply the population mean of response-variate values when the predictor variables are held constant at their means $\bar{x}_1, \bar{x}_2, \ldots, \bar{x}_p$, instead of at zero. Model (127.6) is therefore just the multivariable extension of that for $p = 1$ in (47.3).

In what follows (127.6) will primarily be used; it is a simple matter to incorporate statements to convert the original values x_{ji} into centered values $(x_{ji} - \bar{x}_j)$, as preliminaries for standard computer routines.

Note: It is sometimes convenient to regard the constant terms μ and β_0 as partial regression coefficients of y on a "dummy," uncentered "variable" x_0, which has the same value, $x_0 = 1$, for each observation.

§128. THE MATRIX FORM OF THE REGRESSION MODEL

The response y_i in (127.6) can be written as the product of a row vector of the ith centered variable values and the column vector $[\mu \quad \beta_1 \quad \cdots \quad \beta_p]'$ of the structural parameters, together with the added ε_i, because

$$
\begin{aligned}
y_i &= [1 \quad (x_{1i} - \bar{x}_1) \quad \cdots \quad (x_{pi} - \bar{x}_p)][\mu \quad \beta_1 \quad \cdots \quad \beta_p]' + \varepsilon_i \\
&= \mu + (x_{1i} - \bar{x}_1)\beta_1 + \cdots + (x_{pi} - \bar{x}_p)\beta_p + \varepsilon_i \\
&= \mu + \beta_1(x_{1i} - \bar{x}_1) + \cdots + \beta_p(x_{pi} - \bar{x}_p) + \varepsilon_i
\end{aligned}
$$

This is true, with the same column vector of parameters, for all $i = 1, \ldots, n$, and assembling the individual rows into one matrix allows all the n observations to be collectively modeled by the matrix expression (128.1).

$$
\begin{bmatrix} y_1 \\ \vdots \\ y_i \\ \vdots \\ y_n \end{bmatrix}
=
\begin{bmatrix}
1 & (x_{11} - \bar{x}_1) & \cdots & (x_{p1} - \bar{x}_p) \\
\vdots & \vdots & & \vdots \\
1 & (x_{1i} - \bar{x}_1) & \cdots & (x_{pi} - \bar{x}_p) \\
\vdots & \vdots & & \vdots \\
1 & (x_{1n} - \bar{x}_1) & \cdots & (x_{pn} - \bar{x}_p)
\end{bmatrix}
\begin{bmatrix} \mu \\ \beta_1 \\ \vdots \\ \beta_i \\ \vdots \\ \beta_p \end{bmatrix}
+
\begin{bmatrix} \varepsilon_1 \\ \vdots \\ \varepsilon_i \\ \vdots \\ \varepsilon_n \end{bmatrix}
\tag{128.1}
$$

$$
[\mathbf{Y}] = \qquad\qquad [\mathbf{X}] \qquad\qquad * [\boldsymbol{\beta}] + [\boldsymbol{\varepsilon}] \tag{128.2}
$$

for which, in the abbreviated form (128.2), $\mathbf{Y}$ is the column vector $[y_1 \quad \cdots \quad y_n]'$ on the left of (128.1), $\boldsymbol{\varepsilon}$ is the column vector $[\varepsilon_1 \quad \cdots \quad \varepsilon_n]'$ of the residuals, and $[\boldsymbol{\beta}] = [\mu \quad \beta_1 \quad \cdots \quad \beta_p]'$ is the column vector of parameters which is premultiplied by the $n \times (p+1)$

matrix of predictor variable values, referred to as "the **X**-matrix" or "the model matrix," defined in (128.3).

$$\underset{n \times (p+1)}{\mathbf{X}} = \begin{bmatrix} 1 & (x_{11} - \bar{x}_1) & \cdots & (x_{p1} - \bar{x}_p) \\ \vdots & \vdots & & \vdots \\ 1 & (x_{1n} - \bar{x}_1) & \cdots & (x_{pn} - \bar{x}_p) \end{bmatrix} \tag{128.3}$$

The expression (127.5) can similarly be modeled by taking the successive rows in the model matrix as $[1 \quad x_{1i} \quad \cdots \quad x_{pi}]$, $i = 1, \ldots, n$, and $\boldsymbol{\beta} = [\beta_0 \quad \beta_1 \quad \cdots \quad \beta_p]'$ for the column vector of parameters.

With appropriate identifications we can therefore model the whole data set as

$$\mathbf{Y} = \mathbf{X}\boldsymbol{\beta} + \boldsymbol{\varepsilon}, \qquad \varepsilon_i \sim NI(0, \sigma^2) \tag{128.4}$$

Examples:

(i) For the §52 example where $n = 5$, $p = 1$, and $\bar{x}_1 = 1.44$ so that the centered predictor variable values are $(x_{11} - \bar{x}_1) = 0.86 - 1.44 = -0.58$, $(x_{12} - \bar{x}_1) = -0.41$, and so on, the matrix model for the data set is

$$\begin{bmatrix} 1.38 \\ 1.58 \\ 2.05 \\ 3.01 \\ 4.21 \end{bmatrix} = \begin{bmatrix} 1 & -0.58 \\ 1 & -0.41 \\ 1 & -0.19 \\ 1 & 0.30 \\ 1 & 0.87 \end{bmatrix} \begin{bmatrix} \mu \\ \beta \end{bmatrix} + \begin{bmatrix} \varepsilon_1 \\ \varepsilon_2 \\ \varepsilon_3 \\ \varepsilon_4 \\ \varepsilon_5 \end{bmatrix}$$

$$\begin{array}{cccccc} \mathbf{Y} & = & \mathbf{X} & * \boldsymbol{\beta} & + & \boldsymbol{\varepsilon} \\ (5 \times 1) & & (5 \times 2) & (2 \times 1) & & (5 \times 1) \end{array}$$

(ii) Suppose that the response y is a measure of urban air pollution which is to be related to values of predictor variables, $x_1 =$ urban population and $x_2 =$ urban area, and suppose that the observations (y, x_1, x_2) from $n = 7$ areas were, in coded units, $(1.7, 1.5, 8)$, $(14.4, 6.9, 11.2)$, $(5.5, 5.0, 6.0)$, $(27.3, 6.6, 17.7)$, $(9.8, 1.6, 13.9)$, $(3.5, 2.0, 7.7)$, and $(7.9, 4.4, 5.5)$. Here $p = 2$, $\bar{x}_1 = (1.5 + 6.9 + \cdots + 4.4)/7 = 4.0$, and $\bar{x}_2 = 10.0$ to give the centered values in the last two columns of the 7×3 model matrix in (128.5)

$$\begin{bmatrix} 1.7 \\ 14.4 \\ 5.5 \\ 27.3 \\ 9.8 \\ 3.5 \\ 7.9 \end{bmatrix} = \begin{bmatrix} 1 & -2.5 & -2.0 \\ 1 & 2.9 & 1.2 \\ 1 & 1.0 & -4.0 \\ 1 & 2.6 & 7.7 \\ 1 & -2.4 & 3.9 \\ 1 & -2.0 & -2.3 \\ 1 & 0.4 & -4.5 \end{bmatrix} \begin{bmatrix} \mu \\ \beta_1 \\ \beta_2 \end{bmatrix} + \begin{bmatrix} \varepsilon_1 \\ \vdots \\ \varepsilon_7 \end{bmatrix} \tag{128.5}$$

from which, for example, the model for the first response is generated, into the (127.6) form, as

$$1.7 = y_1 = [1 \quad -2.5 \quad -2.0][\mu \quad \beta_1 \quad \beta_2]' + \varepsilon_1 = \mu - 2.5\beta_1 - 2.0\beta_2 + \varepsilon_1$$

§129. THE PARAMETER ESTIMATION EQUATIONS, THE PARTIAL REGRESSION COEFFICIENT ESTIMATES, AND THE PREDICTION EQUATION

If the estimates of the parameters are $\hat{\mu} = m$, $\hat{\beta}_1 = b_1, \ldots, \hat{\beta}_p = b_p$, conveniently collected as the column vector $\mathbf{b} = [m \quad b_1 \quad \cdots \quad b_p]'$, the predicted value for y at $x_1 = x_{1i}, \ldots, x_p = x_{pi}$, that is, the estimate of $\mu(y \mid \mathbf{X}_i)$ in (127.4), is, extending (48.2),

$$\hat{y}_i = m + b_1(x_{1i} - \bar{x}_1) + \cdots + b_p(x_{pi} - \bar{x}_p)$$

$$= [1 \quad (x_{1i} - \bar{x}_1) \quad \cdots \quad (x_{pi} - \bar{x}_p)]\mathbf{b} \tag{129.1}$$

There will be one such product for each $\hat{y}_i$; the column vector $\mathbf{b}$ is the same for each and the row vectors are the successive row vectors of the model matrix $\mathbf{X}$ in (128.3). Hence, with $\hat{\mathbf{Y}}$ as the column vector $[\hat{y}_1 \quad \cdots \quad \hat{y}_n]'$ of predicted values, we can write

$$\hat{\mathbf{Y}} = \mathbf{Xb} \tag{129.2}$$

and the column vector of the residuals $e_i = (y_i - \hat{y}_i)$, $i = 1, \ldots, n$, is then $\mathbf{e} = [\mathbf{Y} - \hat{\mathbf{Y}}]$.

The sum of squares, S say, of these residuals is, as per §121, Examples (ii) and (iii),

$$S = \sum_1^n e_i^2 = \mathbf{e'e} = [\mathbf{Y} - \hat{\mathbf{Y}}]'[\mathbf{Y} - \hat{\mathbf{Y}}] \tag{129.3}$$

The parameter estimation equations: Theory has shown that the particular estimates of the parameters which make S as small as possible can be obtained by solving the *parameter estimation equations* which are, in matrix form,

$$[\mathbf{X'X}]\mathbf{b} = \mathbf{X'Y} \tag{129.4}$$

Except when p is small, computer programs are used to find the inverse $[\mathbf{X'X}]^{-1}$, and hence the solutions as

$$[\mathbf{X'X}]^{-1}[\mathbf{X'X}]\mathbf{b} = [\mathbf{X'X}]^{-1}\mathbf{X'Y}$$

that is,

$$\mathbf{Ib} = \mathbf{b} = [\mathbf{X'X}]^{-1}\mathbf{X'Y} \tag{129.5}$$

The prediction equation giving the estimate of $\mu(y|\mathbf{X})$ for general values, $x_1, x_2, \ldots, x_p$, of the predictor variables can then be written, using the individual elements of $\mathbf{b}$, as

$$\hat{y} = \bar{y} + b_1(x_1 - \bar{x}_1) + \cdots + b_p(x_p - \bar{x}_p) \tag{129.6}$$

$$= \left(\bar{y} - \sum_1^p b_i\bar{x}_i\right) + b_1 x_1 + \cdots + b_p x_p \tag{129.7}$$

The equation defines a "plane" in $(p+1)$ dimensions.

Notes: (i) Prediction equations should only be used with considerable caution to estimate or predict $\mu(y|\mathbf{X})$ for predictor variable values outside the region of those used to estimate $\boldsymbol{\beta}$ unless other information is available that the model is valid over the wider region.

(ii) As evident in the following examples, the symmetric matrix $\mathbf{X}'\mathbf{X}$ embodies some useful statistics; its 1,1th element is n, the number of observations, and, as a nice consequence of using the model (127.6) with $\mathbf{X}$ from (128.3), $\mathbf{X}'\mathbf{X}$ is a bordered matrix, its other diagonal elements are the corrected sums of squares $\sum' x_j x_j = \sum_i (x_{ji} - \bar{x}_j)^2$, while the off-diagonal elements are the corrected sums of products $\sum' x_j x_k = \sum (x_{ji} - \bar{x}_j)(x_{ki} - \bar{x}_k)$ of pairs of the predictor variables. If the model (127.5) is used, the first row of $\mathbf{X}'\mathbf{X}$ is $[n \quad \sum x_1 \quad \cdots \quad \sum x_p]$ and the j,kth elements are the uncorrected sums of products $\sum_i x_j x_k$, and squares when $j = k$.

(iii) The matrices $\mathbf{X}'\mathbf{Y}$ for the models (127.6) and (127.5) are, respectively,

$$\mathbf{X}'\mathbf{Y} = \left[\sum y \quad \sum{}' x_1 y \quad \cdots \quad \sum{}' x_p y\right],$$

$$\mathbf{X}'\mathbf{Y} = \left[\sum y \quad \sum x_1 y \quad \cdots \quad \sum x_p y\right] \tag{129.8}$$

so that appropriate elements may be selected from the former and $\mathbf{X}'\mathbf{X}$ to calculate simple linear regression coefficients.

(iv) Because the use of the centered predictor variable model (127.6) "automatically" gives $\hat{\mu} = \bar{y}$, it is really only necessary to invert a $p \times p$ matrix to obtain the estimates of the parameters $\beta_1, \ldots, \beta_p$. The full $(p+1) \times (p+1)$ $\mathbf{X}'\mathbf{X}$ matrix is retained here to parallel procedures for the (127.5) model. For this model the prediction equation would be obtained directly, with $\mathbf{b} = [b_0 \, b_1 \ldots b_p]$ from (129.5), as

$$y = b_0 + b_1 x_1 + \cdots + b_p x_p \tag{129.9}$$

wherein $b_0 = (\bar{y} - \sum_1^p b_i\bar{x}_i)$ as in (129.7).

Examples:

(i) With n observations (x_i, y_i) in $p = 1$ regression, the matrices $\mathbf{X}$, $\mathbf{X'Y}$, and $\mathbf{X'X}$ are

$$\mathbf{X} = \begin{bmatrix} 1 & (x_1 - \bar{x}) \\ \vdots & \vdots \\ 1 & (x_n - \bar{x}) \end{bmatrix}, \quad \mathbf{X'Y} = \begin{bmatrix} 1 & \cdots & 1 \\ (x_1 - \bar{x}) & \cdots & (x_n - \bar{x}) \end{bmatrix} \begin{bmatrix} y_1 \\ \vdots \\ y_n \end{bmatrix} = \begin{bmatrix} \sum y_i \\ \sum (x_i - \bar{x}) y_i \end{bmatrix}$$

$$\mathbf{X'X} = \begin{bmatrix} 1 & \cdots & 1 \\ (x_1 - \bar{x}) & \cdots & (x_n - \bar{x}) \end{bmatrix} \begin{bmatrix} 1 & (x_1 - \bar{x}) \\ \vdots & \vdots \\ 1 & (x_n - \bar{x}) \end{bmatrix} = \begin{bmatrix} n & 0 \\ 0 & \sum' xx \end{bmatrix}$$

so that, finding $[\mathbf{X'X}]^{-1}$ from §125 (ii), the estimates of the parameters are, from (129.5),

$$\mathbf{b} = \begin{bmatrix} m \\ b \end{bmatrix} = \begin{bmatrix} 1/n . & 0 \\ 0 & 1/\sum' xx \end{bmatrix} \begin{bmatrix} \sum y_i \\ \sum (x_i - \bar{x}) y_i \end{bmatrix}$$

that is,

$$m = \hat{\mu} = \frac{1}{n} \sum y_i = \bar{y}, \qquad b = \frac{\sum (x_i - \bar{x}) y_i}{\sum' xx} = \frac{\sum' xy}{\sum' xx}$$

exactly as in §48.

(ii) Continuing from (128.5) it can be calculated that

$$\mathbf{X'X} = \begin{bmatrix} 7 & 0 & 0 \\ 0 & 32.34 & 17.94 \\ 0 & 17.94 & 121.48 \end{bmatrix}, \quad \mathbf{X'Y} = \begin{bmatrix} 70.1 \\ 86.63 \\ 196.71 \end{bmatrix} \qquad (129.10)$$

Since $\mathbf{X'X}$ is a bordered matrix, the first step is to find, using §125 (i),

$$\begin{bmatrix} 32.34 & 17.94 \\ 17.94 & 121.48 \end{bmatrix}^{-1} = \begin{bmatrix} 0.033681 & -0.004974 \\ -0.004974 & 0.008966 \end{bmatrix}$$

The complete inverse is then found using §125 (iii) and postmultiplied by $\mathbf{X'Y}$ to give the parameter estimates as

$$\mathbf{b} = \begin{bmatrix} m = \bar{y} \\ b_1 \\ b_2 \end{bmatrix}$$

$$= \begin{bmatrix} 0.142857 & 0 & 0 \\ 0 & 0.033681 & -0.004974 \\ 0 & -0.004974 & 0.008966 \end{bmatrix} \begin{bmatrix} 70.1 \\ 86.63 \\ 196.71 \end{bmatrix} = \begin{bmatrix} 10.0143 \\ 1.9393 \\ 1.3329 \end{bmatrix} \quad (129.11)$$

$$[\mathbf{X'X}]^{-1} \qquad\qquad * \; [\mathbf{X'Y}] \;\; = \qquad \mathbf{b}$$

The individual predicted values, from $\hat{\mathbf{Y}} = \mathbf{Xb}$, and the residuals $\mathbf{e} = [\mathbf{Y} - \hat{\mathbf{Y}}]$ are

$$
\hat{\mathbf{Y}} = \begin{bmatrix} 1 & -2.5 & -2.0 \\ 1 & 2.9 & 1.2 \\ 1 & 1.0 & -4.0 \\ 1 & 2.6 & 7.7 \\ 1 & -2.4 & 3.9 \\ 1 & -2.0 & -2.3 \\ 1 & 0.4 & -4.5 \end{bmatrix} \begin{bmatrix} 10.0143 \\ 1.9393 \\ 1.3329 \end{bmatrix} = \begin{bmatrix} 2.50 \\ 17.24 \\ 6.62 \\ 25.32 \\ 10.56 \\ 3.07 \\ 4.79 \end{bmatrix}, \quad \mathbf{Y} - \hat{\mathbf{Y}} = \begin{bmatrix} -0.80 \\ -2.84 \\ -1.12 \\ 1.98 \\ -0.76 \\ 0.43 \\ 3.11 \end{bmatrix} (129.12)
$$

The residuals are conveniently examined at this stage for early symptoms of model inadequacy (§138). Those above add to zero apart from rounding error and show no obvious patterns when plotted against the corresponding $\hat{y}_i$'s for example.

The prediction equation giving the estimate of $\mu(y|\mathbf{X})$ for any pair of (x_1, x_2) values is, with $\bar{x}_1 = 4$ and $\bar{x}_2 = 10$,

$$
\hat{y} = [1 \quad (x_1 - 4) \quad (x_2 - 10)][10.0143 \quad 1.9393 \quad 1.3329]'
$$

or, in terms of the uncentered predictor variables themselves,

$$
\hat{y} = -11.07 + 1.94x_1 + 1.33x_2
$$

which represents a plane in the $(\hat{y}, x_1, x_2)$ dimensions. The positive partial regression coefficient of y on x_1 suggests that, for the same area, air pollution increases with population number (more cars) and, similarly, $b_2 = 1.33$ suggests that, for the same population size, pollution increases with urban area (more driving being required; perhaps "small *is* beautiful").

§130. THE MULTIPLE REGRESSION ANALYSIS OF VARIANCE

A slightly different form of the ANOVA identity than that analogous to (52.1) is often used in multiple regression analyses. The identity states that the

$$
\text{Total (uncorrected) sum of squares,} \sum_1^n y_i^2 = \mathbf{Y}'\mathbf{Y}
$$

minus the

$$
\text{Sum of squares due to the model parameters,} R(\boldsymbol{\beta})
$$

equals the

$$
\text{Residual sum of squares,} \sum_1^n e_i^2 = \mathbf{e}'\mathbf{e} = [\mathbf{Y} - \hat{\mathbf{Y}}]'[\mathbf{Y} - \hat{\mathbf{Y}}]
$$

Abbreviations used for the multiple regression ANOVA sum of squares "due to," or "accounted for," or "explained by" the parameters and the associated predictor-variable values in the model are

$$R(\boldsymbol{\beta}) = R(\beta_0, \beta_1, \ldots, \beta_p) = R(\mathbf{X}) = R(1, x_1, \ldots, x_p)$$

for the model (127.5), and

$$R(\boldsymbol{\beta}) = R(\mu, \beta_1, \ldots, \beta_p) = R(1, x_1 - \bar{x}_1, \ldots, x_p - \bar{x}_p)$$

for the centered model (127.6).

A theoretical result states that

$$R(\boldsymbol{\beta}) = \mathbf{b}'\mathbf{X}'\mathbf{Y} \qquad (130.1)$$

where $\mathbf{b} = [\mathbf{X}'\mathbf{X}]^{-1}\mathbf{X}'\mathbf{Y}$ from (129.5), and the predictor-variable values in $\mathbf{X}$ are uncentered and centered for models (127.5) and (127.6), respectively. For the centered predictor variables model, the estimate of μ is just $\bar{y}$ so that, with $\mathbf{X}'\mathbf{Y}$ from (129.8),

$$R(\boldsymbol{\beta}) = R(\mu, \beta_1, \ldots, \beta_p) = [\bar{y} \quad b_1 \quad \cdots \quad b_p]$$
$$\times \left[\sum y \quad \sum' x_1 y \quad \cdots \quad \sum' x_p y \right]'$$
$$= \frac{(\sum y)^2}{n} + b_1 \sum' x_1 y + \cdots + b_p \sum' x_p y \qquad (130.2)$$

The first term in (130.2), $\bar{y} \sum y = (\sum y)^2/n$, is the usual correction term. It is also $R(\mu)$, the sum of squares due to the parameter μ, when the simple model $y_i = \mu + \varepsilon_i$ is fitted. The remainder on the right-hand side of (130.2) is the *additional sum of squares* accounted for by the predictor variables when the simple model is extended to include the regression coefficients $\beta_1, \ldots, \beta_p$. Reading "|" as "after fitting," we have that the sum of squares due to the actual predictor parameters per se is denoted by

$$R(\beta_1, \ldots, \beta_p \mid \mu) = b_1 \sum' x_1 y + \cdots + b_p \sum' x_p y \qquad (130.3)$$

and calculated as

$$R(\beta_1, \ldots, \beta_p \mid \mu) = R(\mu, \beta_1, \ldots, \beta_p) - R(\mu) \qquad (130.4)$$
$$= \mathbf{b}'\mathbf{X}'\mathbf{Y} - \frac{(\sum y)^2}{n} \qquad (130.5)$$

It can readily be checked using (129.8) and $b_0 = (\bar{y} - \sum_i b_i \bar{x}_i)$ that, on the basis of the uncentered model (127.5), the sum of squares due to the regression coefficients per se,

$$R(\beta_1, \ldots, \beta_p \mid \beta_0) = R(\beta_0, \beta_1, \ldots, \beta_p) - R(\beta_0)$$

also gives, as it should, exactly the quantity on the right-hand side of (130.3).

Degrees of freedom: Attributing n degrees of freedom to the total sum of squares $\sum y_i^2$, and subtracting one degree of freedom for each parameter fitted, leaves $(n - 1 - p)$ degrees of freedom for the residual sum of squares. Additionally on this, the residual sum of squares $\sum e_i^2$ is distributionally equivalent to the sum of the squares of f, say, independent, unconstrained ε^2's where f is the number n of e_i's less the number of constraints on, or equations interconnecting, them. That the number of constraining equations is $(p + 1)$, there being one constraint for each parameter fitted, can be seen from the parameter estimation equations because, from (129.4)

$$\mathbf{0} = \mathbf{X'Y} - \mathbf{X'Xb} = \mathbf{X'}[\mathbf{Y} - \mathbf{Xb}] = \mathbf{X'e}$$

This gives as many equations in the e_i's as there are rows in $\mathbf{X'}$, that is, $(1 + p)$ and hence again, $\sum e_i^2 = \mathbf{e'e}$ has $(n - 1 - p)$ df. It may be noted, for example, that, since the elements in the first row of $\mathbf{X'}$ are all unity, the first equation in the e_i's is $[1 \quad 1 \quad \cdots \quad 1]\mathbf{e} = 0$; that is, $\sum_1^n e_i = e_1 + \cdots + e_n = 0$.

The complete ANOVA, in which the total sum of squares is partitioned successively downwards, is given in Table 130.1.

Table 130.1. The Multiple Regression ANOVA

Source	df	ss	ms
Total	n	$\mathbf{Y'Y} = \sum y_i^2$	
$R(\mu)$: (due to μ)	1	$\dfrac{(\sum y)^2}{n}$	$\dfrac{(\sum y)^2}{n}$
Total $- R(\mu)$	$n - 1$	$\sum' yy = \sum y^2 - \dfrac{(\sum y)^2}{n}$	s_y^2
$R(\beta_1, \ldots, \beta_p \mid \mu)$:	p	$\mathbf{b'X'Y} - \dfrac{(\sum y)^2}{n}$	s_R^2
$\mathbf{e'e} = [\mathbf{Y} - \hat{\mathbf{Y}}]'[\mathbf{Y} - \hat{\mathbf{Y}}]$:	$(n - 1 - p)$	$\mathbf{Y'Y} - \mathbf{b'X'Y}$	s^2

Example: Continuing from §129 Example (ii), we first find

$$\mathbf{Y'Y} = 1.7^2 + \cdots + 7.9^2 = 1156.49, \quad 7 \text{ df}$$

Second, we find $R(\mu)$, the correction term, as

$$R(\mu) = \bar{y} \sum y = \frac{(\sum y)^2}{n} = \frac{70.1^2}{7} = 702.00, \quad 1 \text{ df}$$

Third, we find $\mathbf{b'X'Y} = R(\mu, \beta_1, \beta_2)$, the sum of squares accounted for by the three parameters μ, β_1, and β_2 as

$$R(\mu, \beta_1, \beta_2) = [10.0143 \quad 1.9393 \quad 1.3329][70.1 \quad 86.63 \quad 196.71]'$$
$$= 1132.20, \quad 3 \text{ df}$$

Subtractions then give

(i)   Total $- R(\mu) = 1156.49 - 702.00 \quad = 454.49, \quad 6 \text{ df}$

(ii) $R(\beta_1, \beta_2 \mid \mu) = 1132.20 - 702.00 \quad = 430.20 \quad 2 \text{ df}$

(iii) Residual ss $= \mathbf{Y'Y} - R(\mu, \beta_1, \beta_2) = \text{total} - R(\mu) - R(\beta_1, \beta_2 \mid \mu)$

$$= 454.49 - 430.20 \quad = 24.29, \quad 4 \text{ df}$$

and this, of course, is equal to the sum of squares $\sum e_i^2 = \mathbf{e'e} = [\mathbf{Y} - \hat{\mathbf{Y}}]'[\mathbf{Y} - \hat{\mathbf{Y}}]$, of the residuals in (129.12).

The calculations can now be arranged to give the ANOVA in Table 130.2.

Table 130.2. A Multiple Regression ANOVA Example

Source	df	ss	ms	F_T
$\sum y^2$	7	1156.49		
$R(\mu) = \dfrac{(\sum y)^2}{n}$	1	702.00	702.00	9.27[a]
$\sum' yy$	6	454.49	75.75	
Predictors	2	430.20	215.10	35.44[b]
Residual, $\mathbf{e'e}$	4	24.29	6.07	

[a] $P < 0.025$.
[b] $P < 0.005$.

§131. HYPOTHESIS TESTS AND ESTIMATIONS FOR THE REGRESSION COEFFICIENTS

The ANOVA provides two immediate hypotheses tests. The first is simply that for (17.1) with $\mu_T = 0$ for which $F_T = (\sum y)^2/ns_y^2$ for comparison with $F = F(1, n-1; \alpha)$. In effect, this test merely examines whether μ is zero or not in the simplest model, $y_i = \mu + \varepsilon_i$; this test does *not* examine whether the constant term is zero or not in the full model which *includes* the predictor parameters. For the Table 130.2 example, $F_T = 9.27 > F_\alpha = F(1, 6; 0.05) = 5.99$.

The second hypothesis test is that of

$$H_T: \beta_1 = \cdots = \beta_p = 0, \quad H_A: \text{not all the } \beta_j \text{ are zero}; \qquad \alpha \quad (131.1)$$

The test statistic, from Table 130.1, is $F_T = s_R^2/s^2$ for comparison with $F_\alpha = F(p, n-1-p; \alpha)$. If $F_T < F_\alpha$ the implication is that the model including the prediction variables gives no significant improvement over the primitive model, $y = \mu + \varepsilon$, as a data explanation. Nor, if $F_T > F_\alpha$ does acceptance of H_A imply that every one of the β_j is nonzero but, less specifically, that one or more of the $\beta_j \neq 0$. Some procedures for detecting possibly redundant regression coefficients are discussed in §138.

Example: For the data in the Table 130.2 example, where $p = 2$, the hypothesis specification (131.1) is

$$H_T: \beta_1 = \beta_2 = 0, \quad H_A: \beta_1 \text{ or } \beta_2 \text{ or both} \neq 0; \qquad \alpha$$

From Table 130.2, $F_T = 35.44$ and, since $F(2, 4; 0.005) = 26.284 < F_T$, the acceptance of H_A is strongly indicated with an exceedance probability of less than 0.005.

Individual partial regression coefficients: To examine a partial regression coefficient, the result from §132[+] is needed that:

The estimate of the variance of the partial regression coefficient b_j is the ANOVA residual mean square s^2, multiplied by the diagonal element $c^{(j+1)(j+1)}$, in the $(j+1)$th (*not* the jth) row and column of $[\mathbf{X'X}]^{-1}$; that is,

$$\hat{V}(b_j) = s^2 c^{(j+1)(j+1)}, \qquad j = 0, 1, \ldots, p \qquad (131.2)$$

where, for $j = 0$, $b_0 = \hat{\beta}_0$ for the uncentered model, and $b_0 = \hat{\mu} = \bar{y}$ for the centered model, it being noted that $[\mathbf{X'X}]^{-1}$ will differ for the two models.

Accordingly, to examine the hypothesis that the jth population partial regression coefficient is equal to some specified test value β_{jT}, say,

$$H_T: \beta_j = \beta_{jT}, \quad H_A: \beta_j \neq \beta_{jT}: \quad \alpha$$

The test statistic is

$$F_T = \frac{(b_j - \beta_{jT})^2}{s^2 c^{(j+1)(j+1)}} \tag{131.3}$$

and H_A will be accepted if F_T falls in the critical region of F-variate values exceeding $F_\alpha = F(1, n-1-p; \alpha)$.

The $100(1-\alpha)\%$ confidence interval estimate of β_j is

$$b_j \mp \sqrt{F_\alpha s^2 c^{(j+1)(j+1)}}, \quad j = 0, 1, \ldots, p \tag{131.4}$$

Examples: Continuing from §130, to test

(i) $$H_T: \mu = 0, \quad H_A: \mu \neq 0$$

we take the first diagonal element of $[\mathbf{X}'\mathbf{X}]^{-1}$ in (129.11) and s^2 from Table (130.2) to obtain the test statistic value

$$F_T = \frac{(\bar{y} - 0)^2}{s^2 c^{11}} = \frac{(10.0143)^2}{(6.07)(0.142857)} = 115.7$$

which exceeds $F_{0.001} = F(1, 4; 0.001) = 74.137$ to give strong support for H_A.

The 99% confidence interval estimate for μ is, with $F(1, 4; 0.01) = 21.20$,

$$10.0143 \mp \sqrt{(21.20)(6.07)(0.142857)} = 5.73, 14.30$$

(ii) Taking, for example, the one-sided alternative hypothesis test of

$$H_T: \beta_1 = 1, \quad H_A: \beta_1 > 1; \quad 0.05$$

we refer $\sqrt{F_T}$ from (131.3) to a $t(n-1-p)$-distribution. The test statistic, with $c^{22} = 0.033681$, is

$$t_T = \frac{1.9393 - 1}{\sqrt{(6.07)(0.033681)}} = 2.08$$

which is close to, but does not quite exceed, $t(4; 0.05) = 2.132$. The 95% confidence interval estimate for β_1 is, from (131.4), or using $t(4; 0.025) = \sqrt{F(1, 4; 0.05)} = 2.776$,

$$1.9393 \mp 2.776\sqrt{(6.07)(0.033681)} = 0.68, 3.19$$

Presentation of results: Some uses of multiple regression estimates are described in later sections. When the main interest is in the equation itself, a common presentation would comprise the prediction equation, the estimated standard deviations of the partial regression coefficient estimates, and the value of the adjusted coefficient of determination R_A^2 (§137). For the example here, with standard deviations $s(b_j)$ from (131.2), we can accordingly offer the following: The prediction equation is $\hat{y} = -11.07 + 1.94x_1 + 1.33x_2$; $R_A^2 = 0.92$; β_1 is estimated as $b_1 = 1.94$, $s(b_1) = 0.45$; $\hat{\beta}_2 = b_2 = 1.33$, $s(b_2) = 0.23$.

§132⁺. VARIANCES AND COVARIANCES OF THE PARTIAL REGRESSION COEFFICIENT ESTIMATES

The elements of the column × row matrix $\varepsilon\varepsilon'$, which appeared as the last example in §121, are variates and the population mean $\mu(\varepsilon\varepsilon')$ of the matrix $\varepsilon\varepsilon'$ is obtained by taking the population means of the individual elements in their respective places. The population means of the diagonal elements are all the same, $\mu(\varepsilon_1^2) = \cdots = \mu(\varepsilon_n^2) = \sigma^2$, from the definition of variance (§5) with $\mu(\varepsilon_i) = 0$. And, on the assumption that the ε_i's are independent, the population means of products of different ε's are zero, from (57.4) with $\rho = 0$. In all, therefore, writing $\mathbf{I}_n = \mathbf{I}$,

$$\mu(\varepsilon\varepsilon') = \mathbf{I}\sigma^2 \tag{132.1}$$

The regression coefficients themselves are, from (129.5), given by $\mathbf{b} = [\mathbf{X'X}]^{-1}\mathbf{X'Y}$ so that, with the model equation $\mathbf{Y} = \mathbf{X}\boldsymbol{\beta} + \boldsymbol{\varepsilon}$,

$$\mathbf{b} = [\mathbf{X'X}]^{-1}\mathbf{X'}[\mathbf{X}\boldsymbol{\beta} + \boldsymbol{\varepsilon}] = [\mathbf{X'X}]^{-1}[\mathbf{X'X}]\boldsymbol{\beta} + [\mathbf{X'X}]^{-1}\mathbf{X'}\boldsymbol{\varepsilon}$$

$$= \mathbf{I}\boldsymbol{\beta} + [\mathbf{X'X}]^{-1}\mathbf{X'}\boldsymbol{\varepsilon}$$

$$= \boldsymbol{\beta} + [\mathbf{X'X}]^{-1}\mathbf{X'}\boldsymbol{\varepsilon} \tag{132.2}$$

If the last matrix in (132.2) were multiplied out it would show that each individual estimate b_j is equal to its population mean β_j plus a linear expression in the ε_i's. Since this has population mean zero, $\mu(b_j) = \beta_j$; that is, b_j is an unbiased estimate of β_j. Furthermore, since the same ε_i's appear in the expression for each of the b_j's, pairs of the b_j's will generally be correlated and not independent (see also §68⁺).

The diagonal elements of the column × row matrix product $[\mathbf{b} - \boldsymbol{\beta}][\mathbf{b} - \boldsymbol{\beta}]'$ are $(\bar{y} - \mu)^2$, $(b_1 - \beta_1)^2, \ldots, (b_p - \beta_p)^2$ of which the population means are therefore $V(\bar{y})$, $V(b_1), \ldots, V(b_p)$. For the off-diagonal elements, taking for example, the 2, 3th = the 3, 2th element: the product $(b_1 -$

$\beta_1)(b_2 - \beta_2)$ has population mean

$$\mu\{(b_1 - \beta_1)(b_2 - \beta_2)\} = CV(b_1, b_2)$$

which is the population covariance of b_1 and b_2 by the definition (57.5). The population mean of $[\mathbf{b} - \boldsymbol{\beta}][\mathbf{b} - \boldsymbol{\beta}]'$ will therefore generate all the variances and covariances required. Hence, with $[\mathbf{b} - \boldsymbol{\beta}]$ from (132.2), and the product transposition rule in §122 which gives $[\mathbf{b} - \boldsymbol{\beta}]' = \boldsymbol{\varepsilon}'\mathbf{X}[\mathbf{X}'\mathbf{X}]^{-1}$, because $[\mathbf{X}'\mathbf{X}]^{-1}$ is a symmetric matrix, we find

$$[\mathbf{b} - \boldsymbol{\beta}][\mathbf{b} - \boldsymbol{\beta}]' = [\mathbf{X}'\mathbf{X}]^{-1}\mathbf{X}'\boldsymbol{\varepsilon}\boldsymbol{\varepsilon}'\mathbf{X}[\mathbf{X}'\mathbf{X}]^{-1}$$

Taking the x-values as constants, the population mean of this is obtained by replacing $\boldsymbol{\varepsilon}\boldsymbol{\varepsilon}'$ by its population mean $\mathbf{I}\sigma^2$ from (132.1). It follows that

$$\mu\{[\mathbf{b} - \boldsymbol{\beta}][\mathbf{b} - \boldsymbol{\beta}]'\} = [\mathbf{X}'\mathbf{X}]^{-1}\mathbf{X}'\mathbf{I}\sigma^2\mathbf{X}[\mathbf{X}'\mathbf{X}]^{-1}$$
$$= \sigma^2[\mathbf{X}'\mathbf{X}]^{-1}[\mathbf{X}'\mathbf{X}][\mathbf{X}'\mathbf{X}]^{-1} = \sigma^2[\mathbf{X}'\mathbf{X}]^{-1} \quad (132.3)$$

because $[\mathbf{X}'\mathbf{X}]^{-1}[\mathbf{X}'\mathbf{X}] = \mathbf{I}$.

One consequence of this result when the analysis is based on the centered model (127.6) is that the elements in the first row of $[\mathbf{X}'\mathbf{X}]^{-1}$, except the 1, 1th element, are zero, as in the §129 examples. This means that $\hat{\mu} = \bar{y}$ is, with normality, independent of each of the remaining regression coefficients $b_1, \ldots, b_p$. If the uncentered variables model (127.5) is used, b_0 is the estimate of β_0 and the elements in the first row of $s^2[\mathbf{X}'\mathbf{X}]^{-1}$ give the estimates $\hat{V}(b_0)$ and $\widehat{CV}(b_0, b_j)$.

§133. HYPOTHESIS TESTS AND INTERVAL ESTIMATES FOR A y-POPULATION MEAN AT A SET OF SPECIFIED x-PREDICTOR VALUES

The mean of the conceptual population of y-values at a set of specified x-values, $x_1 = x_{1s}, \ldots, x_p = x_{ps}$, where these may, but need not, be any one of the observation sets, is

$$\mu[y \mid \mathbf{X}_s] = \mu + \beta_1(x_{1s} - \bar{x}_1) + \cdots + \beta_p(x_{ps} - \bar{x}_p) \quad (133.1)$$

Its point estimate, from (129.1), with $m = \bar{y}$, is

$$\hat{\mu}[y \mid \mathbf{X}_s] = \hat{y}_s = \bar{y} + b_1(x_{1s} - \bar{x}_1) + \cdots + b_p(x_{ps} - \bar{x}_p) \quad (133.2)$$

With $\mathbf{X}'_s = [1 \ (x_{1s} - \bar{x}_1) \ \cdots \ (x_{ps} - \bar{x}_p)]$, $\mu(y \mid \mathbf{X}_s)$ and $\hat{y}_s$ are seen as the matrix products,

$$\mu[y \mid \mathbf{X}_s] = \mathbf{X}'_s\boldsymbol{\beta} \quad \text{and} \quad \hat{y}_s = \mathbf{X}'_s\mathbf{b} \quad (133.3)$$

and a result from (134.5) with $l = \hat{y}_s$ and $\mathbf{C} = \mathbf{X}_s$ is that

$$V(\hat{y}_s) = \mathbf{X}'_s[\mathbf{X}'\mathbf{X}]^{-1}\mathbf{X}_s\sigma^2 \quad \text{so that} \quad \hat{V}(\hat{y}_s) = \mathbf{X}'_s[\mathbf{X}'\mathbf{X}]^{-1}\mathbf{X}_s s^2 \quad (133.4)$$

where s^2 is the residual mean square in the ANOVA.

Hypothesis testing: It follows from (133.4) that, if μ_{sT} is a test value for $\mu(y \mid \mathbf{X}_s)$, the test statistic for

$$H_T: \mu(y \mid \mathbf{X}_s) = \mu_{sT}, \qquad H_A: \mu(y \mid \mathbf{X}_s) \neq \mu_{sT}; \qquad \alpha$$

is, for comparison with $F_\alpha = F(1, n - 1 - p; \alpha)$,

$$F_T = \frac{(\hat{y}_s - \mu_{sT})^2}{s^2 \mathbf{X}'_s[\mathbf{X}'\mathbf{X}]^{-1}\mathbf{X}_s} \quad (133.5)$$

The $100(1 - \alpha)\%$ confidence interval for $\mu(y \mid \mathbf{X}_s)$ is

$$\hat{y}_s \mp \sqrt{F_\alpha s^2 \mathbf{X}'_s[\mathbf{X}'\mathbf{X}]^{-1}\mathbf{X}_s} \quad (133.6)$$

which, if $\mu_{sT} = 0$ and the alternative hypothesis is two-sided, can also be calculated as

$$\hat{y}_s \left(1 \mp \sqrt{\frac{F_\alpha}{F_T}}\right) \quad (133.7)$$

Two estimations of common practical interest are that for the overall mean μ and that for the intercept $\beta_0 = \mu - \sum_1^P \beta_i \bar{x}_i$. These can be examined by taking $\mathbf{X}'_s = [1 \quad 0 \quad \cdots \quad 0]$ and $\mathbf{X}'_s = [1 \quad -\bar{x}_1 \quad \cdots \quad -\bar{x}_p]$, respectively. As in the $p = 1$ case, the narrowest $100(1 - \alpha)\%$ confidence intervals are those for the prediction of μ.

Examples:

(i) Continuing the $p = 1$ case from §129, Example (i), where

$$\mathbf{b} = \begin{bmatrix} \bar{y} \\ b \end{bmatrix} \quad \text{and} \quad [\mathbf{X}'\mathbf{X}]^{-1} = \begin{bmatrix} 1/n & 0 \\ 0 & 1/\sum' xx \end{bmatrix}$$

we find, with $\mathbf{X}'_s = [1 \quad (x_s - \bar{x})]$, the point estimate of $\mu(y \mid \mathbf{X}_s)$ from (133.3) as

$$\hat{y}_s = [1 \quad (x_s - \bar{x})]\begin{bmatrix} \bar{y} \\ b \end{bmatrix} = \bar{y} + b(x_s - \bar{x})$$

Furthermore,

$$\mathbf{X}'_s[\mathbf{X}'\mathbf{X}]^{-1}\mathbf{X}_s = [1 \quad (x_s - \bar{x})]\begin{bmatrix} 1/n & 0 \\ 0 & 1/\sum' xx \end{bmatrix}\begin{bmatrix} 1 \\ x_s - \bar{x} \end{bmatrix}$$

$$= [1 \quad (x_s - \bar{x})]\begin{bmatrix} 1/n \\ (x_s - \bar{x})/\sum' xx \end{bmatrix} = \frac{1}{n} + \frac{(x_s - \bar{x})^2}{\sum' xx}$$

Hence, since $\sqrt{F(1, n-1-p; \alpha)} = t(n-1-p; \alpha/2)$, taking $p = 1$ and $s^2 = s^2_{y|x}$, (133.6) gives exactly the (54.4) expression for the $100(1-\alpha)\%$ confidence interval for $\mu(y|X_s)$.

(ii) Continuing from §129, Example (ii), suppose it is required to examine the population mean of y-values at $x_1 = 5$, $x_2 = 8$ so that, since $\bar{x}_1 = 4.0$, $\bar{x}_2 = 10.0$, $X'_s = [1 \quad (5-4) \quad (8-10)] = [1 \quad 1 \quad -2]$.

With the values of b and $[X'X]^{-1}$ from (129.11), the point estimate of $\mu(y|X_s)$ is, from (133.3),

$$\hat{y}_s = [1 \quad 1 \quad -2][10.0143 \quad 1.9393 \quad 1.3329]' = 9.29$$

and we find, using rounded elements of $[X'X]^{-1}$,

$$X'_s[X'X]^{-1}X_s = [1 \quad 1 \quad -2]\begin{bmatrix} 0.1429 & 0 & 0 \\ 0 & 0.0337 & -0.005 \\ 0 & -0.005 & 0.009 \end{bmatrix}\begin{bmatrix} 1 \\ 1 \\ -2 \end{bmatrix} = 0.2326$$

Hence, with $s^2 = 6.07$ (4 df) from the ANOVA, (133.6) gives the 95% confidence interval for $\mu(y|X_s)$

$$9.29 \mp \sqrt{(7.709)(6.07)(0.2326)} = 6.0, 12.6$$

The value 10, for example, lies in this interval so that the inference from the above hypothesis test with $\mu_{sT} = 10$ would be the nonrejection of H_T.

§134⁺. LINEAR COMBINATIONS OF PARAMETERS AND OF ESTIMATES

If $c_0, c_1, \ldots, c_p$ are constant numbers and $\bar{y}, b_1, \ldots, b_p$ are the estimates of $\mu, \beta_1, \ldots, \beta_p$, the quantity

$$l = c_0\bar{y} + c_1b_1 + \cdots + c_pb_p \tag{134.1}$$

is a general linear combination of the parameter estimates. In matrix form, with $b = \beta + [X'X]^{-1}X'\varepsilon$ from (132.2) and writing

$$C = [c_0 \quad c_1 \quad \cdots \quad c_p]', \qquad b = [\bar{y} \quad b_1 \quad \cdots \quad b_p]' \tag{134.2}$$

it follows that

$$l = C'b = C'\beta + C'[X'X]^{-1}X'\varepsilon \tag{134.3}$$

and that, since $\mu[\varepsilon] = [0 \quad 0 \quad \cdots \quad 0]'$,

$$\mu[l] = \lambda = C'\beta = c_0\mu + c_1\beta_1 + \cdots + c_p\beta_p \tag{134.4}$$

Practically interesting combinations of the parameters can now be generated by suitable choices of the c-constants and examined using the corresponding estimates.

Examples:

(i) The predicted value $\mu[y\,|\,\mathbf{X}_s]$ in (133.3) is obtained from (134.4) with $\mathbf{C}' = \mathbf{X}'_s$ and its estimate is then, from (134.1), just

$$l = \mathbf{C}'\mathbf{b} = \mathbf{X}'_s\mathbf{b} = \hat{y}_s$$

(ii) Particular combinations of the partial regression coefficients can be examined. Thus,

$$\lambda = \beta_1 - 2\beta_2 + \beta_3$$

is generated from (134.4) by taking

$$\mathbf{C}' = [0 \quad 1 \quad -2 \quad 1 \quad 0 \quad \cdots \quad 0]$$

and estimated from (134.3) as

$$l = \mathbf{C}'\mathbf{b} = b_1 - 2b_2 + b_3$$

Variance:

The variance of l in (134.3) is $V(l) = \mu[(l - \lambda)^2]$ where, because $(l - \lambda)$ is a scalar, $(l - \lambda)^2 = [l - \lambda][l - \lambda]'$. Now, with $(\mathbf{b} - \boldsymbol{\beta}) = [\mathbf{X}'\mathbf{X}]^{-1}\mathbf{X}'\boldsymbol{\varepsilon}$ from (132.2),

$$(l - \lambda) = \mathbf{C}'\mathbf{b} - \mathbf{C}'\boldsymbol{\beta} = \mathbf{C}'(\mathbf{b} - \boldsymbol{\beta}) = \mathbf{C}'[\mathbf{X}'\mathbf{X}]^{-1}\mathbf{X}'\boldsymbol{\varepsilon}$$

so that using the matrix product transposition rule for $[l - \lambda]'$,

$$(l - \lambda)^2 = \mathbf{C}'[\mathbf{X}'\mathbf{X}]^{-1}\mathbf{X}'\boldsymbol{\varepsilon}\boldsymbol{\varepsilon}'\mathbf{X}[\mathbf{X}'\mathbf{X}]^{-1}\mathbf{C}$$

The population mean of this quantity is obtained by replacing $\boldsymbol{\varepsilon}\boldsymbol{\varepsilon}'$ by its population mean $\mathbf{I}\sigma^2$ from (132.1) to give $V(l)$ as

$$V(l) = \mathbf{C}'[\mathbf{X}'\mathbf{X}]^{-1}\mathbf{X}'\mathbf{I}\sigma^2\mathbf{X}[\mathbf{X}'\mathbf{X}]^{-1}\mathbf{C}$$

$$= \mathbf{C}'[\mathbf{X}'\mathbf{X}]^{-1}[\mathbf{X}'\mathbf{X}][\mathbf{X}'\mathbf{X}]^{-1}\mathbf{C}\sigma^2 = \mathbf{C}'[\mathbf{X}'\mathbf{X}]^{-1}\mathbf{C}\sigma^2 \qquad (134.5)$$

of which the estimate $\hat{V}(l)$ is obtained by replacing σ^2 by s^2 from the ANOVA.

Furthermore, it was shown in §132 that $[\mathbf{X}'\mathbf{X}]^{-1}\sigma^2 = \mu\{[\mathbf{b} - \boldsymbol{\beta}][\mathbf{b} - \boldsymbol{\beta}]'\}$ is a matrix of which the elements are the variances and covariances between pairs of the regression coefficient estimates. Writing $[\mathbf{CV}(\mathbf{b})]$ for this matrix, (134.5) can also be written

$$V(l) = \mathbf{C}'[\mathbf{CV}(\mathbf{b})]\mathbf{C} \qquad (134.6)$$

Examples:

(i) Using the particular vector C from Example (ii) above, it can be checked that (134.6) generates $V(l) = V(b_1 - 2b_2 + b_3)$ as

$$V(b_1 - 2b_2 + b_3) = V(b_1) + 4V(b_2) + V(b_3) - 4CV(b_1, b_2) + 2CV(b_1, b_3)$$
$$- 4CV(b_2, b_3)$$

The hypothesis test of

$$H_T: \beta_1 - 2\beta_2 + \beta_3 = 0, \quad H_A: \beta_1 - 2\beta_2 + \beta_3 \neq 0; \quad \alpha$$

can be carried out as an ordinary t-test with

$$t_T = \frac{b_1 - 2b_2 + b_3}{\sqrt{\hat{V}(b_1 - 2b_2 + b_3)}}$$

where the denominator is calculated from (134.5) with s^2 $(n - 1 - p \text{ df})$ from the ANOVA instead of σ^2.

(ii) Continuing from §129, Example (ii), and Table 130.2 it may be checked that for the test of

$$H_T: \beta_1 - 2\beta_2 = 0, \quad H_A: \beta_1 - 2\beta_2 \neq 0; \quad 0.05$$

taking $C' = \begin{bmatrix} 0 & 1 & -2 \end{bmatrix}$ gives the test statistic value

$$t_T = \frac{(b_1 - 2b_2)}{\sqrt{s^2 C'[X'X]^{-1}C}} = \frac{-0.7263}{\sqrt{(6.07)(0.0897)}} = -0.98$$

giving, because $|t_T| = 1.004 < t(4; 0.025) = 2.776$, no evidence for rejecting H_T.

Note: The above derivations are given primarily to support the most common applications, those in §133, it being opined that other linear combinations of the partial regression coefficients may not be of great practical interest unless the corresponding predictor variables have the same dimensions; one example however, is, described in §136, Note (ii).

§135. THE PREDICTION INTERVAL FOR ANOTHER, "FUTURE" OBSERVATION

For this situation, defined in the $p = 1$ case in §55, the arguments and the procedure described there extend exactly to $p \geq 1$ situations. If y_s is the "future" observation to be obtained at x-values $x_1 = x_{1s}, \ldots, x_p = x_{ps}$, and $X'_s = \begin{bmatrix} 1 & (x_{1s} - \bar{x}_1) & \cdots & (x_{ps} - \bar{x}_p) \end{bmatrix}$, the $100(1 - \alpha)\%$ prediction interval is

$$\hat{y}_s \mp \sqrt{F_\alpha\{s^2 + \hat{V}(\hat{y}_s)\}} = \hat{y}_s \mp \sqrt{F_\alpha s^2\{1 + \mathbf{X}_s'[\mathbf{X}'\mathbf{X}]^{-1}\mathbf{X}_s\}} \qquad (135.1)$$

where $\hat{y}_s = \mathbf{X}_s'\mathbf{b}$ and $F_\alpha = F(1, n-1-p; \alpha)$.

Example: Continuing from §133, Example (ii), it may be checked that the 95% prediction interval for a "future" observation at $x_1 = 5$ and $x_2 = 8$ is from 1.7 to 16.9, appreciably wider than the 95% confidence interval found for the y-population mean at these (x_1, x_2)-values.

§136. POLYNOMIAL REGRESSION: CURVE FITTING

Suppose that n pairs of observations (x_i, y_i), $i = 1, \ldots, n$, have been obtained and, in extension of the straight line procedures, §46ff, it is required to fit a least squares curve, of degree p to describe the data, based on the structural relationship

$$\mu(y \mid x_i) = \beta_0 + \beta_1 x_i + \beta_2 x_i^2 + \cdots + \beta_p x_i^p \qquad (136.1)$$

This is seen to be analogous to a multiple regression relationship with $x_1 = x$, $x_2 = x^2, \ldots, x_p = x^p$ except that the previous interpretations of the β's as partial regression coefficients no longer obtain—it being unrealistic to consider x^2, for example, as constant while allowing x to vary. Instead, the maximum nonzero coefficient indicates the degree of the polynomial curve in just the two dimensions, x and y. Thus, if $\beta_2 \neq 0$ for $p = 2$, the curve is quadratic; $\beta_3 \neq 0$ for $p = 3$ indicates the curve as cubic and so on.

The perfect, though not linear, relationships between pairs $x_j = x^j$ and $x_k = x^k$ can cause special difficulties, sometimes obscurely referred to as multicollinearity, when $\mathbf{X}'\mathbf{X}$ is inverted unless large numbers of decimal places are used. Nor are these difficulties removed if the centered variables $(x_j - \bar{x}_j)$ and $(x_k - \bar{x}_k)$ are used, since subtraction of constants does not change correlation. The point has been recently discussed, specifically in the present context, by Bradley and Srivastava (1979) whose results confirm advantages for the simplest intuitively suggested palliative: we couch the relationship in term of $(x - \bar{x})$ and its successive powers. Accordingly, with $y_i = \mu(y \mid x_i) + \varepsilon_i$, the model used is

$$y_i = \beta_0 + \beta_1(x - \bar{x}) + \beta_2(x - \bar{x})^2 + \cdots + \beta_p(x - \bar{x})^p + \varepsilon_i$$

$$\varepsilon_i \sim NI(0, \sigma^2), \qquad i = 1, \ldots, n \qquad (136.2)$$

The multiple regression procedures now applicable are those for the

(127.5) form of the model with the predictor variables $x_1 = (x - \bar{x}), \ldots, x_p = (x - \bar{x})^p$.

Example: In an article on aging, *The Des Moines Register* (25 July 1982) published the data in Table 136.1 to show the diminution with age of the thickness of individual hairs of males. It being clear that the decline is nonlinear, we will fit a $p = 2$ quadratic curve to the data using the computationally more convenient quantities $x = (age/10)$ in decades and $y = (hair\ thickness - 94)\ \mu m$ to give the predictor variables and their y-variate correspondents in Table 136.2.

Table 136.1. Individual Hair Thickness (μm) and Age (years)

Age	20	30	40	50	60	70
Hair thickness	101	98	96	94	86	80

Table 136.2. $x =$ Age (decades), $y =$ Hair Thickness $- 94$ (μm)

x	$x_1 = (x - \bar{x})$	$x_2 = (x - \bar{x})^2$	y	$\hat{y}$	$(y - \hat{y})$
2	−2.5	6.25	7	6.4	0.6
3	−1.5	2.25	4	5.1	−1.1
4	−0.5	0.25	2	2.4	−0.4
5	0.5	0.25	0	−1.7	1.7
6	1.5	2.25	−8	−7.2	−0.8
7	2.5	6.25	−14	−14.0	0.0
Total 27	0	17.5	−9	−9	0

The model matrix **X** for (136.2) has unit elements in the first column and columns 2 and 3 of Table 136.2 are, respectively, columns 2 and 3 of **X**. Hence,

$$\mathbf{X'X} = \begin{bmatrix} 6 & 0 & 17.5 \\ 0 & 17.5 & 0 \\ 17.5 & 0 & 88.375 \end{bmatrix} \quad \text{and} \quad \mathbf{X'Y} = \begin{bmatrix} -9 \\ -71.5 \\ -52.25 \end{bmatrix}$$

it being noted that the 2, 3th element of $\mathbf{X'X}$, $\sum x_1 x_2$, will not be zero unless, as here, the x-values are symmetric about their mean. Inversions of general $\mathbf{X'X}$ matrices are part of the computer routines commonly used for multiple regression analyses, especially for large p. Here a computer obtained $\mathbf{X'X}$ as that giving the regression coefficient estimates:

$$\mathbf{b} = \begin{bmatrix} b_0 \\ b_1 \\ b_2 \end{bmatrix} = [\mathbf{X'X}]^{-1}\mathbf{X'Y} = \begin{bmatrix} 0.394531 & 0 & -0.078125 \\ 0 & 0.057143 & 0 \\ -0.078125 & 0 & 0.026786 \end{bmatrix} \begin{bmatrix} -9 \\ -71.5 \\ -52.25 \end{bmatrix}$$

$$= \begin{bmatrix} 0.5313 \\ -4.0857 \\ -0.6963 \end{bmatrix}$$

and hence the estimation or prediction equation is

$$\hat{y} = 0.5313 - 4.0857(x - \bar{x}) - 0.6964(x - \bar{x})^2 \qquad (136.3)$$

which, with $(x - \bar{x})^2 = x^2 - 2x\bar{x} + \bar{x}^2$ and $\bar{x} = 4.5$, reduces to

$$\hat{y} = 4.8149 + 2.1819x - 0.6904x^2 \qquad (136.4)$$

Either equation may be used to obtain the (rounded) values for $\hat{y}$ and the residuals $(y - \hat{y})$ in Table 136.2. The residuals are not alarming. The prediction equation may be presented in terms of the original data with age in years as, from (136.4),

$$\text{Hair thickness} = 98.8149 + 2.1819\left(\frac{\text{age}}{10}\right) - 0.6964\left(\frac{\text{age}^2}{100}\right)$$

$$= 98.8149 + 0.21819(\text{age}) - 0.006964(\text{age})^2 \qquad (136.5)$$

The ANOVA, like that in Tables 130.1 and 130.2, is given in Table 136.3. The ANOVA test statistic $F_T = 88.39$ shows that $H_T: \beta_1 = \beta_2 = 0$ is decisively rejected. For the individual regression coefficients, the test

Table 136.3. ANOVA for Table 136.2

Source	df	ss	ms	F_T
Total	6	329		
$R(\mu)$	1	13.5		
$\sum' yy$	5	315.5	63.1	
$R(\beta_1, \beta_2 \mid \mu)$	2	310.236	155.118	88.39[a]
Residual	3	5.264	1.755	

[a]$P < 0.005$.

statistics for H_T: $\beta_j = 0$, H_A: $\beta_j \neq 0$, $j = 1, 2$, are from (131.3), $F_{T1} = (4.0857)^2/(1.755)(0.057) = 166.9$ and $F_{T2} = 0.6964^2/(1.755)(0.026) = 10.6$ which, compared with critical values from the $F(1, 3)$-distribution, suggest that $\beta_1 \neq 0$ and $\beta_2 \neq 0$ with exceedance probabilities $P \approx 0.001$ and $P < 0.05$, respectively. Confidence intervals for the regression coefficients, for predicted values at designated x-values and prediction intervals may also be obtained using the multiple regression procedures previously described.

Notes: (i) Although the above example illustrates a generally applicable curve fitting procedure, the particular feature of the data here, that the x-values are equally spaced, admits an alternative, simpler, analysis. This uses tabulated values for orthogonal polynomial, contrast coefficients to obtain immediate estimates of the regression coefficients because $X'X$ is then diagonal. For details see, for example, Dowdy and Wearden (1983).

(ii) Another possibility of interest here is the examination of the regression coefficients when the prediction equation is re-expressed in terms of the original x-variables rather than centered, or orthogonal polynomial, functions of them. Thus, the coefficient of x in (136.4) is in fact obtained as

$$b_1 - (2\bar{x})b_2 = -4.0857 + 2\bar{x}(0.6964) = 2.1819$$

To test the hypothesis that the regression coefficient estimated by 2.1819 is zero or not, the procedure in §134$^+$ with $\mathbf{C}' = [0 \quad 1 \quad -2\bar{x}]$ gives, with $\bar{x} = 4.5$ and $s^2 = 1.755$, the test statistic

$$t_T = 2.1819/\sqrt{1.755\{0.057 + 4(4.5^2)(0.027) - 4(4.5)(0)\}} = 1.10$$

which, compared with critical regions from the $t(3)$-distribution, suggests a zero regression coefficient and the reduction of the model to become just $y = \alpha_0 + \alpha_1 x^2 + \varepsilon$. Estimates of α_0 and α_1 can be found by simple linear regression procedures, using x^2 as the one predictor variable, to give the more parsimonious prediction equations:

$$\hat{y} = 9.1611 - 0.4602x^2 \quad \text{and} \quad \text{hair thickness} = 103.16 - 0.0046(\text{age})^2$$

with $s^2_{y|x} = 1.84$ (4 df) and a good-looking set of residuals for the former.

(iii) The example nicely illustrates the fact that multiple regression models, in the absence of theoretical substantiation, are locally descriptive rather than veridical representations of the data-generating processes. Thus, (136.5) suggests that those living beyond 140 years may expect negative hair thickness, a phenomenon with little trichological substantiation to date, even from Bulgaria whence such longevity has been reported.

§137. R^2: THE COEFFICIENT OF DETERMINATION

The efficacy of the p predictor variables in reducing the unexplained residual variability in the predicted variable can be indexed by the coefficient of determination R^2, $0 \leq R^2 \leq 1$, defined as that fraction of the corrected total sum of squares which is accounted for by the inclusion of the partial regression coefficients in the model. Accordingly,

$$R^2 = \frac{R(\beta_1, \ldots, \beta_p \mid \mu)}{\sum' yy} = \frac{\mathbf{b'X'Y} - R(\mu)}{\mathbf{Y'Y} - R(\mu)} \qquad (137.1)$$

and R^2 is the $p \geq 1$ extension of the $p = 1$ coefficient in (52.3). That R^2 provides a measure of goodness of fit also follows from the fact that the square of the correlation coefficient between y_i and $\hat{y}_i$ is R^2 as previously noted for $p = 1$.

Examples: For the urban air pollution data, R^2 is obtained from the ANOVA, Table 130.2, as $R^2 = 430.20/454.49 = 0.95$. For the §136 data, the ANOVA table shows that $R^2 = 310.236/315.5 = 0.98$, indicating a very good fit.

The test statistic for the hypothesis specification that all the regression coefficients are zero (131.1) can be expressed in terms of R^2. For this, from (137.1), $R(\beta_1, \ldots, \beta_p \mid \mu) = R^2 \sum' yy$ so that the residual sum of squares $\sum' yy - R(\beta_1, \ldots, \beta \mid \mu) = (1 - R^2) \sum' yy$. Hence, the test statistic (compare the ANOVA, Table 130.1) is

$$F_T = \frac{s_R^2}{s^2} = \frac{R^2 \sum' yy/p}{(1 - R^2) \sum' yy/(n - 1 - p)}$$

$$= \frac{(n - 1 - p)R^2}{p(1 - R^2)} \qquad (137.2)$$

Models with different numbers of prediction variables for the same data are sometimes compared (§138). R^2 is then not completely satisfactory as a comparison index because it necessarily increases with p, for constant $\sum' yy$, because $R(\beta_1, \ldots, \beta_p \mid \mu)$ so increases. An alternative index, related to R^2 but which does not necessarily increase with p, is the adjusted or shrunken R^2-coefficient

$$R_A^2 = 1 - \frac{s^2}{s_y^2} \tag{137.3}$$

where $s_y^2 = \sum' yy/(n-1)$ as in Table 130.1.

Examples: From the respective ANOVA, this index takes the values
$1 - (6.07/75.75) = 0.92$ and $1 - (1.755/63.1) = 0.97$ for the above two
examples.

§138. EXAMINING MODEL ADEQUACY

Model inadequacy can arise in several ways because, in particular, either
or both of (i) the assumptions made about the ε-component and (ii) the
specification of the structural part of the model, may be incorrect. The
latter can be a particularly debilitating deficiency because if the structural
part is under specified, that is, has too few predictor variables, parametric
contamination of the residual component by the effects of the unspecified
regression parameters may induce autocorrelated residual elements and
make it difficult to define the underspecification. Some research on this is
reported in Ponder (1982).

Residuals: As noted in §109, the residuals $e_i = (y_i - \hat{y}_i)$ may be examined
for information on model adequacy. Texts, such as Draper and Smith
(1981), describe a half-normal plotting procedure that uses residuals to
assess the validity of the normal distribution assumption.

In multiple regression, plots of e_i may be made against each predictor
variable in turn. If the plot of e_i against the values of an x_j shows a
pattern such as a progression of negative, positive, and negative again
residuals, it suggests that a linear term alone in x_j is not sufficient to
describe the jth predictor effect; further predictor variables, in
$x_j^2, x_j^3, \ldots$, might then improve the fit. Neither should the plot of e_i
against $\hat{y}_i$ show any trend; a trend may suggest either misspecification or
that $V(\varepsilon_i)$ is not constant. A data conversion (§110) may possibly
alleviate the latter problem. More information on residuals in multiple
regression situations is available in Chatterjee and Price (1977).

Which predictor variables? The problem of which variables to include in
a model has progressed in importance from individual early discussions
(e.g., Cox, 1958) to become an area of much research activity. The
general problem is that models may be underspecified, by including too

few predictor variables, or overspecified, by including redundant predictor variables. Several variable selection procedures have been recommended such as the Mallows C_p-index procedure which was examined, and gave good results, in the study by Bendel and Afifi (1977). These authors also address a vexatious related problem on just what α-levels are appropriate when successive, related, significance tests are made. Forward selection procedures are attractive if predictor variables can be ordered in decreasing relative importance as in some polynomial regression situations where the lowest powers of x may be the most influential in a restricted x-range. Otherwise, backward selection procedures appear advantageous as being less subject to parametric contamination of the successive residual mean squares. Such stepwise regression procedures are authoritatively described in Draper and Smith (1981). In general, since with p predictor variables each of the parameters $\mu, \beta_1, \ldots, \beta_p$ may be zero or nonzero, there are (2^{p+1}) possible models. In the absence of extraneous information on the relative importances of individual or sets of predictor variables, a complete resolution would involve the examination of all the models and there may be little to choose between some of them especially if high correlations exist between several predictor variables. Therefore, when p is large, computer algorithms such as that in Hocking and Leslie (1967) are indicated; see also Furnival (1971). When p is small, the procedure described next can be used to explore the variable selection aspect of model adequacy.

Suppose that out of p possible predictor variables, the stage has been reached that none of $\mu, \beta_1, \beta_2, \ldots, \beta_r$, $r < p$, is zero and that a further q parameters, $\beta_{r+1}, \ldots, \beta_{r+q}$, are being considered as candidates for inclusion in the model. In many cases q is just one. The choice therefore lies between the two models

$$y = \mu + \sum_{j=1}^{r} \beta_j(x_j - \bar{x}_j) + \varepsilon \quad \text{and} \quad y = \mu + \sum_{j=1}^{r+q} \beta_j(x_j - \bar{x}_j) + \varepsilon$$

and the question is "Does the inclusion of the additional parameters $\beta_{r+1}, \ldots, \beta_{r+q}$, *after* the $\beta_1, \ldots, \beta_r$, improve the model?" One way to assess improvement involves the choice between the hypotheses:

$$H_T: \beta_j \neq 0 \text{ for } j = 1, \ldots, r \quad \text{and} \quad \beta_j = 0 \text{ for } j = r+1, \ldots, r+q$$

and

$$H_A: \beta_j \neq 0 \text{ for } j = 1, \ldots, r \quad \text{and} \quad \text{one or more of } \beta_j \neq 0$$
$$\text{for } j = r+1, \ldots, r+q \qquad (138.1)$$

The appropriate test statistic is

$$F_T = \frac{R(\beta_{r+1}, \ldots, \beta_{r+q} \mid \mu, \beta_1, \ldots, \beta_r)}{qs^2}$$

$$= \frac{R(\beta_1, \ldots, \beta_{r+q} \mid \mu) - R(\beta_1, \ldots, \beta_r \mid \mu)}{qs^2} \qquad (138.2)$$

where s^2, $(n-1-r-q$ df), is obtained from the ANOVA carried out, with $p = r + q$ now, as in Table 130.1, and we decide in favor of H_T or H_A if, respectively, F_T is less than or greater than $F_\alpha = F(q, n-1-r-q; \alpha)$.

If all but one of the whole set of p predictor variables have been fitted, (138.2) may be used to assess the importance of the remaining variable. In SAS PROC GLM, for example, the output TYPE III SS column gives the numerators of (138.2), $R(\beta_1 \mid \mu, \beta_2, \ldots, \beta_p), \ldots, R(\beta_p \mid \mu, \beta_1, \ldots, \beta_{p-1})$, for the alternative sets of $(p-1)$ initially fitted variables. The output TYPE I SS gives $R(\beta_1 \mid \mu), R(\beta_2 \mid \mu, \beta_1), \ldots, R(\beta_p \mid \mu, \beta_1, \ldots, \beta_{p-1})$ for the forward sequential examination of the individual predictor variables taken in the data input order.

Notes: (i) If all but one, x_j say, of p variables have been included in the model, the above test for the additional inclusion of x_j is equivalent to the test of H_T: $\beta_j = 0$, H_A: $\beta_j \neq 0$, exemplified in §131. (ii) Accurate prediction, rather than good, retrospective, data description is important for a good model. It is accordingly generally judicious to include marginally significant predictors rather than to strive for a model having as few parameters as possible, especially if s^2 is generously endowed with degrees of freedom.

Example: Continuing the urban air pollution example, H_T: $\mu = 0$ was rejected in §131 and the models remaining for consideration are:

(a) $y = \mu + \beta_1(x_1 - \bar{x}_1) + \varepsilon$.
(b) $y = \mu + \beta_2(x_2 - \bar{x}_2) + \varepsilon$.
(c) $y = \mu + \beta_1(x_1 - x_1) + \beta_2(x_2 - \bar{x}_2) + \varepsilon$.

The first model could be adopted if the additional inclusion of the x_2-predictor, to obtain model (c), gives no improvement. Similarly, model (b) could be adopted if including β_1 *after* β_2 has been fitted gives no improvement. Beginning with $\sum' yy = $ total ss $- R(\mu)$ from Table 130.2, the calculations can be conveniently arranged as in Table 138, it being noted that this is an $r = 1$ and $q = 1$ situation.

Table 138. ANOVA for Testing Model Adequacy

Source	(i) df	ss	ms	F_T	Source	(ii) df	ss	ms	F_T
$\sum' yy$	6	454.49			$\sum' yy$	6	454.49		
$R(\beta_1, 0 \mid \mu)$	1	232.06	232.06	5.22	$R(0, \beta_2 \mid \mu)$	1	318.53	318.53	11.71
Residual	5	222.43	44.49		Residual	5	135.96	27.19	
$R(\beta_2 \mid \mu, \beta_1)$	1	198.14	198.14	32.64	$R(\beta_1 \mid \mu, \beta_2)$	1	111.67	111.67	18.40
Residual	4	24.29	6.07		Residual	4	24.29	6.07	

In analysis (i), $R(\beta_1, 0 \mid \mu) = (\sum' x_1 y)^2 / \sum' x_1 x_1 = 86.63^2/32.34$ is the $p = 1$ sum of squares due to β_1 with model (a) in which $\beta_2 = 0$ and $R(\beta_2 \mid \mu, \beta_1) = R(\beta_1, \beta_2 \mid \mu) - R(\beta_1, 0 \mid \mu) = 430.20 - 232.06$ with $R(\beta_1, \beta_2 \mid \mu) = 430.20$ from Table 130.2. Values in analysis (ii) are similarly obtained. The sums of squares $R(\beta_1, 0 \mid \mu)$ and $R(\beta_2 \mid \mu, \beta_1)$ also appear successively as the two TYPE I SS entries while $R(\beta_1 \mid \mu, \beta_2)$ is the first TYPE III SS entry, in the SAS PROC GLM output.

In analysis (i) the first test statistic $F_T = 5.22$ is less than $F(1, 5; 0.05) = 6.61$ while in (ii) $F_T = 11.71 > F(1, 5; 0.025) = 10.01$ so that, if a model using only one of the predictor variables is required, it would have to be model (b). From (ii), however, we note further that, because the second $F_T = 18.40$ exceeds $F(1, 4; 0.025)$, the addition of β_1 to model (b) does effect a significant improvement. Model (c) wins. Supportingly, with $s_y^2 = 454.49/6 = 75.75$, (137.3) gives the R_A^2-values, 0.41, 0.64, and 0.92, for models (a), (b), and (c), respectively.

Note: Whenever variables are added to or excluded from a model, previously calculated estimates need revision. Formulas are available for this but it may be expected that computer routines will deal with the whole model determination procedure allowing alternative options for model adequacy criteria.

§139. STANDARDIZED PARTIAL REGRESSION COEFFICIENTS

The estimate of the population mean of y observations at the set $(x_1, x_2, \ldots, x_p)$ of predictor variate values is

$$\hat{y} = \bar{y} + b_1(x_1 - \bar{x}_1) + \cdots + b_p(x_p - \bar{x}_p) \tag{139.1}$$

With $s_y^2 = \sum' yy/(n-1)$ and defining $s_j^2 = \sum' x_j x_j/(n-1)$, $j = 1, \ldots, p$, although, with fixed predictor variable models, the s_j^2 are not strictly variance estimates, we can (i) transpose the $\bar{y}$ to the left-hand side, (ii) divide both sides by s_y, and (iii) note that $b_j = b_j s_j/s_j$ and write (139.1) as

$$\frac{\hat{y} - \bar{y}}{s_y} = \left(\frac{b_1 s_1}{s_y}\right)\left(\frac{x_1 - \bar{x}_1}{s_1}\right) + \cdots + \left(\frac{b_p s_p}{s_y}\right)\left(\frac{x_p - \bar{x}_p}{s_p}\right) \qquad (139.2)$$

The standardized partial regression coefficient estimates are the quantities $b_j^* = (b_j s_j/s_y)$, $j = 1, \ldots, p$, in (139.2). The dimensions of s_j/s_y "cancel" those of b_j so that each b_j^* is dimension-free. The quantities $x_j^* = (x_j - \bar{x})/s_j$ are also dimension-free and, with $\hat{y}^* = (\hat{y} - \bar{y})/s_y$, (139.2) gives the standardized estimation or prediction equation as just

$$\hat{y}^* = b_1^* x_1^* + \cdots + b_p^* x_p^* \qquad (139.3)$$

in which b_j^*, for example, measures the amount by which $\hat{y}$ changes in s_y-units for a one s_j, standard deviation-unit change in the predictor variable x_j when the values of all the other predictor variables are constant.

Example: Assembling the statistics $\bar{y} = 10.0143$, $b_1 = 1.9393$, $b_2 = 1.3329$, $s_y^2 = 454.49/6 = 75.75$, and from $\mathbf{X'X}$, $s_1^2 = 32.34/6 = 5.39$, $s_2^2 = 121.48/6 = 20.2467$, we find for the urban pollution example

$$b_1^* = 1.9393 \sqrt{\frac{32.34}{454.49}} = 0.5173, \qquad b_2^* = 1.3329 \sqrt{\frac{121.48}{454.49}} = 0.6891$$

and the standardized estimation equation is

$$\frac{\hat{y} - 10.0143}{8.7034} = \hat{y}^* = 0.5173 x_1^* + 0.6891 x_2^*$$

Comparisons of the now dimensionless standardized partial regression coefficients are more useful than comparisons of the unstandardized coefficients. We see here that although $b_1 > b_2$, $b_1^* < b_2^*$ which reflects the fact that x_2 alone is more effective in "explaining" the y-variability than is x_1 alone. This interpretation is pursued in §140.

§140. PARTITIONING R^2

Although the general result is obtained in §141$^+$, the discussion here is restricted for simplicity to the $p = 2$ case. The result required, from (141.2) with $p = 2$, is that the sum of squares due to the parameters β_1

and β_2, $R(\beta_1, \beta_2 | \mu)$, can alternatively be expressed as

$$R(\beta_1, \beta_2 | \mu) = b_1^2 {\sum}' x_1 x_1 + b_2^2 {\sum}' x_2 x_2 + 2 b_1 b_2 {\sum}' x_1 x_2$$

Hence, the coefficient of determination, (137.1), is

$$R^2 = b_1^2 \left(\frac{\sum' x_1 x_1}{\sum' yy} \right) + b_2^2 \left(\frac{\sum' x_2 x_2}{\sum' yy} \right) + 2 b_1 b_2 \left(\frac{\sum' x_1 x_2}{\sum' yy} \right) \qquad (140.1)$$

Now $\sum' x_1 x_1 / \sum' yy = (n-1) s_1^2 / (n-1) s_y^2 = s_1^2 / s_y^2$ and, similarly, the coefficient of b_2^2 is s_2^2 / s_y^2. Furthermore, if $r_{12} = \sum' x_1 x_2 / (n-1) s_1 s_2$ is the equivalent correlation coefficient between x_1 and x_2, $\sum' x_1 x_2 = (n-1) r_{12} s_1 s_2$. Using these results (140.1) becomes

$$R^2 = \frac{b_1^2 s_1^2}{s_y^2} + \frac{b_2^2 s_2^2}{s_y^2} + \frac{2 b_1 b_2 r_{12} s_1 s_2}{s_y^2}$$

which, with the definition of the standardized partial correlation coefficients in §139, gives

$$R^2 = b_1^{*2} + b_2^{*2} + 2 b_1^* b_2^* r_{12} \qquad (140.2)$$

This shows that the *squares* of the standardized partial regression coefficients measure the contributions to R^2 due to the corresponding individual predictor variables, while the $2 b_1^* b_2^* r_{12}$ term measures their joint contribution. The complement $(1 - R^2)$ measures the relative contribution of the residual, or the "unexplained," variability. Defining $e^* = (e/s_y) = (y - \hat{y})/s_y$, $1 - R^2$ is in fact just $\sum (e^*)^2 / (n-1)$.

Example: Continuing the urban pollution example with $b_1^* = 0.5173$, $b_2^* = 0.6891$, $s_1^2 = 5.39$, $s_2^2 = 20.2467$ from §139 and, from $\mathbf{X}'\mathbf{X}$ in §129, $17.94 = (n-1) r_{12} s_1 s_2 = 6 r_{12} s_1 s_2$ so that $r_{12} = 0.2862$. Hence, the individual contribution of the first predictor variable is

$$b_1^{*2} = 0.5173^2 = 0.268$$

the greater individual contribution of the second predictor variable is

$$b_2^{*2} = 0.6891^2 = 0.475$$

the joint contribution is

$$2 b_1^* b_2^* r_{12} = 2(0.5173)(0.6891)(0.2862) = 0.204$$

the combined total is $0.268 + 0.475 + 0.204 = 0.947 = R^2$, and the relative contribution of the residual variability is $1 - R^2 = 0.053$.

An alternative partitioning of R^2: The sum of squares due to all the predictor variables may be partitioned as

$$R(\beta_1, \ldots, \beta_p \mid \mu) = R_1 + R_2 \qquad (140.3)$$

where $R_1 = R(\beta_1, \ldots, \beta_r, 0, \ldots, 0 \mid \mu)$ is the sum of squares due to the first r predictor variables fitted, the remaining $(p-r)$ variables being ignored, and $R_2 = R(\beta_{r+1}, \ldots, \beta_p \mid \mu, \beta_1, \ldots, \beta_r)$ is the sum of squares due to the last $(p-r)$ variables, after the first r have been fitted. Division by $\sum' yy$ gives

$$R^2 = \frac{R(\beta_1, \ldots, \beta_p \mid \mu)}{\sum' yy} = \frac{R_1}{\sum' yy} + \frac{R_2}{\sum' yy} \qquad (140.4)$$

In this, $R_1/\sum' yy$ is the contribution to R^2 of the first r variables together with that due to commonalities which these variables have with the last $(p-r)$ variables. $R_2/\sum' yy$ measures the orthogonal contribution of the remaining $(p-r)$ variables, that contribution, the commonalities having been allowed for, attributable to the last $(p-r)$ variables alone.

Example: From Table 138 (ii) for first fitting β_2 and then β_1, $R(\beta_2, 0 \mid \mu) = 318.53$, $R(\beta_1 \mid \mu, \beta_2) = 111.67$, and $\sum' yy = 454.49$. Hence, $R_1 = 318.53/454.49 = 0.701$ and $R_2 = 111.67/454.49 = 0.246$ and $0.701 + 0.246 = 0.947 = R^2$.

§141[+]. ALTERNATIVE EXPRESSIONS FOR REGRESSION SUMS OF SQUARES

From (130.1) the sum of squares due to all the parameters is $R(\mu, \beta_1, \ldots, \beta_p) = \mathbf{b}'[\mathbf{X'Y}]$ and $[\mathbf{X'Y}] = \mathbf{X'Xb}$ from (129.4). It follows that

$$R(\mu, \beta_1, \ldots, \beta_p) = \mathbf{b'X'Xb} \qquad (141.1)$$

With $\mathbf{X'X}$ as defined in §129, Note (ii), $\mathbf{b'X'Xb}$ is the product $\mathbf{X'Xb}$ premultiplied by $\mathbf{b}'$. The result for the right-hand side of (141.1) is that

$$\mathbf{b'}[\mathbf{X'X}]\mathbf{b} = n\bar{y}^2 + b_1^2 \sum{}' x_1 x_1 + \cdots + b_p^2 \sum{}' x_p x_p + 2b_1 b_2 \sum{}' x_1 x_2$$

$$+ 2b_1 b_3 \sum{}' x_1 x_3 + \cdots + 2b_{p-1} b_p \sum{}' x_{p-1} x_p \qquad (141.2)$$

Furthermore, since $R(\mu) = \bar{y} \sum y = n\bar{y}^2$ it follows that the sum of squares due to the predictor coefficients in the model, $R(\beta_1, \ldots, \beta_p \mid \mu) = R(\mu, \beta_1, \ldots, \beta_p) - R(\mu)$, consists of the terms on the right-hand side of (141.2) after the first term has been subtracted out.

Example: For $p = 3$, (141.2) gives

$$R(\beta_1, \beta_2, \beta_3 \mid \mu) = b_1^2 \sum{}' x_1 x_1 + b_2^2 \sum{}' x_2 x_2 + b_3^2 \sum{}' x_3 x_3 + 2 b_1 b_2 \sum{}' x_1 x_2$$

$$+ 2 b_1 b_3 \sum{}' x_1 x_3 + 2 b_2 b_3 \sum{}' x_2 x_3$$

Multiple Correlation

§142. CORRELATION CONTEXTS

Whereas regression analysis is essentially concerned with prediction, correlation analysis is concerned with assessing the strength of the association between variates. Two particular differences are:

(i) In multiple regression one variate, y, is singled out as the predicted variable. In multiple correlation studies no one variable is necessarily singled out and the data are regarded as n independent observations where *each observation* consists of a *set* comprised of one value for *each of the variates* on record. Accordingly, with $q \geq 2$, the ith observation and the data are

$$(x_{1i}, x_{2i}, \ldots, x_{qi}) \quad \text{for } i = 1, 2, \ldots, n$$

(ii) Regression predictions can be made whether the predictor variables are fixed or random. In correlation studies, all the x_j, $j = 1, \ldots, q$, are presumed to be random variates. Sample correlation coefficients are then calculated as estimates of population correlation coefficients.

§143. CORRELATION COEFFICIENT TYPES

When $q = 2$, only the simple correlation coefficient (§57) is calculated. For $q > 2$, several types of correlation coefficient can be informatively examined. Some of these are:

(i) **Simple correlation coefficients:** These are correlation coefficients calculated between pairs of variates x_j and x_k, from (59.1) with $x = x_j$

245

and $y = x_k$. And there are $(q-1)q/2$ such coefficients altogether; the ten $r_{12}, r_{13}, r_{14}, r_{15}, r_{23}, r_{24}, r_{25}, r_{34}, r_{35}, r_{45}$ if $q = 5$, for example.

(ii) **Partial correlation coefficients:** The simple correlation coefficient between x_j and x_k ignores the influence which the other variates may jointly have on x_j and x_k. For example, in the simple correlation coefficient r_{12}, between $x_1 = $ weight and $x_2 = $ a physical agility measure, the effects of $x_3 = $ age, on both weight and physical agility, are ignored. The *partial* correlation coefficient $r_{12|3}$ eliminates the interference of age by estimating the correlation between weight and agility supposing that age is *held constant*.

Exemplifying the notation further, if $q \geqq 5$, $r_{12|345}$ denotes the partial correlation coefficient between x_1 and x_2 when the variates x_3, x_4, and x_5 are held constant and all other variates are ignored.

(iii) **Multiple correlation coefficients:** Taking $q = 5$, for example, suppose that the variates can be divided into two naturally distinct sets (a) x_1 and x_2 and (b) x_3, x_4, and x_5. If $c_1, \ldots, c_5$ are constant numbers, two correlated univariates can be calculated as $u = c_1 x_1 + c_2 x_2$ and $v = c_3 x_3 + c_4 x_4 + c_5 x_5$. The simple correlation coefficient r_{uv} is then a kind of multiple correlation coefficient. This topic can be pursued as canonical correlation in texts on multivariate analysis. As a special case, if the sets are (a) one variate only, x_5 say, and (b) the remaining four, so that $u = x_5$ and $v = c_1 x_1 + \cdots + c_4 x_4$, the c-constants that maximize r_{uv}^2 are in fact the respective partial regression coefficients of x_5 on $x_1, \ldots, x_4$. The maximum value of r_{uv}^2 is the coefficient of determination R^2 in (137.1). When all the variates are random, R is termed the multiple correlation coefficient.

§144. ESTIMATING AND TESTING PARTIAL CORRELATION COEFFICIENTS

In general, partial correlation coefficients are conveniently calculated from the inverse, $\mathbf{R}^{-1}$ of the symmetric correlation matrix $\mathbf{R}$ in which, for the variates $x_1, \ldots, x_q$, the j, kth element is the simple correlation coefficient r_{jk} between x_j and x_k so that

$$\mathbf{R} = \begin{bmatrix} 1 & r_{12} & \cdots & r_{1q} \\ r_{21} & 1 & \cdots & r_{2q} \\ \cdot & \cdot & & \cdot \\ \cdot & \cdot & & \cdot \\ r_{q1} & r_{q2} & \cdots & 1 \end{bmatrix} \qquad (144.1)$$

If r^{jk} is the j, kth element of $\mathbf{R}^{-1}$, the partial correlation coefficient between x_j and x_k, the other variates being constant, is written and estimated as

$$r_{jk|(\text{all except } j,k)} = \frac{-r^{jk}}{\sqrt{r^{jj}r^{kk}}} \qquad (144.2)$$

The particular result for $q = 3$ is that

$$r_{jk|l} = \frac{r_{jk} - r_{jl}r_{kl}}{\sqrt{(1 - r_{jl}^2)(1 - r_{kl}^2)}} \qquad (144.3)$$

Example: Taking the urban pollution data with $\mathbf{y}$ as the variable x_3, the elements of $\mathbf{R}$ can be obtained from those in $\mathbf{X'X}$ and $\mathbf{X'Y}$; thus, from §129 and with $\sum' yy = 454.49$ from §130, $r_{13} = 86.63/\sqrt{(32.34)(454.49)} = 0.715$. The complete matrix $\mathbf{R}$ and its inverse were computed to be

$$\mathbf{R} = \begin{bmatrix} 1 & 0.286 & 0.715 \\ 0.286 & 1 & 0.837 \\ 0.715 & 0.837 & 1 \end{bmatrix},$$

$$\mathbf{R}^{-1} = \begin{bmatrix} 6.1452 & 6.4125 & -9.7610 \\ 6.4125 & 10.0310 & -12.9809 \\ -9.7610 & -12.9809 & 18.8441 \end{bmatrix}$$

The point estimate of $r_{23|1}$, for example, is obtained using (144.2) as

$$r_{23|1} = -\frac{(-12.9809)}{\sqrt{(10.0310)(18.8441)}} = 0.944$$

Equivalently, from (144.3) with elements from $\mathbf{R}$,

$$r_{23|1} = \frac{0.837 - (0.286)(0.715)}{\sqrt{\{1 - (0.286^2)\}\{1 - (0.715)^2\}}} = 0.944$$

Hypothesis testing: The hypothesis that a population partial correlation coefficient is zero can be examined by the usual t-test procedure. Thus, writing $\rho_{jk|(\text{all except } j,k)}$ as ρ for brevity here and its estimate $r_{jk|(\text{all except } jk)}$ from (144.2) simply as r, the test statistic for the specification

$$H_T: \rho = 0, \quad H_A: \rho \neq 0; \quad \alpha$$

can be taken as

$$t_T = + \sqrt{\frac{(n-q)r^2}{1-r^2}}$$

and H_T is indicated unless $t_T > t_{\alpha/2} = t(n-q; \alpha/2)$.

Example: Continuing from the above, where $q = 3$, for

$$H_T: \rho_{32|1} = 0, \quad H_A: \rho_{32|1} \neq 0; \qquad 0.01$$

H_A is indicated because

$$t_T = 0.944 \sqrt{\frac{7-3}{1-(0.944)^2}} = 5.72 > t_{0.005} = t(4; 0.005) = 4.604$$

Confidence interval estimation: When n is large, an approximate $100(1-\alpha)\%$ confidence interval for a partial correlation coefficient can be obtained by:

(i) converting the partial correlation coefficient to z' as

$$z' = \ln \sqrt{\frac{1+r}{1-r}}$$

(ii) finding the confidence interval for $\mu(z')$ as

$$\bar{\zeta}_L, \bar{\zeta}_H = z' \mp z_{\alpha/2} \sqrt{\frac{1}{n-q-1}} \tag{144.4}$$

(iii) using the reverse conversion

$$\tilde{\rho} = \frac{e^{2\bar{\zeta}} - 1}{e^{2\bar{\zeta}} + 1}$$

to convert $\bar{\zeta}_L, \bar{\zeta}_H$ into the corresponding interval for $\rho_{jk|(\text{all except } j,k)}$.

The procedure is exactly that described in §65 except that $(n-3)$ in (65.1) is replaced by $(n-q-1)$ to obtain (144.4).

§145. ESTIMATING AND TESTING THE MULTIPLE CORRELATION COEFFICIENT

The multiple correlation coefficient between one variate, taken here as x_1, and the linear combination $\sum_2^q \beta_j x_j$ of the remaining $(q-1)$ variates

can be estimated as $|\sqrt{R^2}|$ from (137.1) by taking $x_1 = y$ and other x_j's as predictor variables. Denoting the population mu relation coefficient as ρ for brevity, $H_T: \rho = 0$ is accepted if F rejected with $P[\text{Type I error}] \leq \alpha$ in favor of $H_A: \rho \neq 0$ if $F_T \geq$

$$F_T = \frac{(n-q)R^2}{(q-1)(1-R^2)} \quad \text{and} \quad F_\alpha = F(q-1, n-q; \alpha)$$

There will be no linear association between x_1 and the variates if $\beta_2, \ldots, \beta_q$ are all zero. Accordingly, the above test and $H_T: \beta_2 = \cdots = \beta_q = 0$ are equivalent so that F_T in (145. $q = p + 1$, just that in (137.2).

References

Angus, J. E. and R. E. Schafer (1984). Improved confidence statements for the binomial parameter, *The American Statistician*, **38**, 189.

Anscombe, F. J. (1948). The transformation of Poisson, binomial and negative binomial data, *Biometrika*, **35**, 246.

Bancroft, T. A. (1968). *Topics in Intermediate Statistical Methods*, Iowa State University Press, Ames, IA.

Bancroft, T. A. and C. P. Han (1981). *Statistical Theory and Inference in Research*, Dekker, New York.

Barnett, V. and T. Lewis (1978). *Outliers in Statistical Data*, Wiley, New York.

Bendel, R. B. and A. A. Afifi (1977). Comparison of stopping rules in forward "stepwise" regression, *Journal of the American Statistical Association*, **72**, 46.

Benson, G. K., A. T. Cowie, C. P. Cox, D. S. Flux, and S. J. Folley (1955). Studies on the hormonal induction of mammary growth and lactation in the goat, *Journal of Endocrinology*, **13**, 46.

Bradley, R. A. and S. S. Srivastava (1979). Correlation in polynomial regression, *The American Statistician*, **33**, 11.

Chatterjee, S. and B. Price (1977). *Regression Analysis by Example*, Wiley, New York.

Clopper, C. J. and E. S. Pearson (1934). The use of confidence or fiducial limits illustrated in the case of the binomial, *Biometrika*, **26**, 404.

Cochran, W. G. (1954). The combination of estimates from different experiments, *Biometrics*, **10**, 101.

Cochran, W. G. and G. M. Cox (1957). *Experimental Designs*, 2nd ed., Wiley, New York.

Conover, W. J. (1980). *Practical Nonparametric Statistics*, 2nd ed., Wiley, New York.

Cox, C. P. (1958). The analysis of Latin Square designs with individual curvatures in one direction, *Journal of the Royal Statistical Society*, B, **20**, 193.

Cox, C. P. (1982). An alternative way of calculating the χ^2 independence or association test statistic for a $2 \times k$ contingency table. *The American Statistician*, **36**, 133.

Cox, C. P. (1985). On comparing the means of two normal populations with variances proportional to the means or to their squares, *Biometrics*, **41**, 261.

Cox, C. P. and T. D. Roseberry (1966). A note on the variance of the distribution of sample number in sequential probability ratio tests, *Technometrics*, **8**, 700.

David, F. N. (1954). *Tables of the Correlation Coefficient*, Biometrika Office, London.

Dowdy, S. and S. Wearden (1983). *Statistics for Research*, Wiley, New York.

Draper, N. R. and H. Smith (1981). *Applied Regression Analysis*, 2nd ed., Wiley, New York.

Dunnett, C. W. (1955). A multiple comparisons procedure for comparing several treatments with a control, *Journal of the American Statistical Association*, **50**, 1096.

Dunnett, C. W. (1964). New tables for multiple comparisons with a control, *Biometrics*, **20**, 482.

Ehrenburg, A. S. C. (1981). The problem of numeracy, *The American Statistician*, **35**, 67.

Ellis, E. L. and M. Delbruck (1939). The growth of bacteriophage, *Journal of General Physiology*, **22**, 365.

Fienberg, S. E. (1977). *The Analysis of Cross-Classified Categorical Data*, The MIT Press, Cambridge, MA.

Furnival, G. M. (1971). All possible regressions with less computation, *Technometrics*, **13**, 403.

Gibbons, J. D. and J. W. Pratt (1975). P-values: Interpretation and methodology, *The American Statistician*, **29**, 20.

Good, I. J. (1983). The calculation of X^2 for two-rowed contingency tables, *The American Statistician*, **37**, 94.

Hocking, R. R. and R. N. Leslie (1967). Selection of the best sub-set in regression, *Technometrics*, **9**, 531.

Hoel, P. G. (1962). *Introduction to Mathematical Statistics*, 3rd ed., Wiley, New York.

Hollander, M. and D. A. Wolfe (1973). *Nonparametric Statistical Methods*, Wiley, New York.

Iman, R. L. (1982). Graphs for use with the Lilliefors test for normal and exponential distributions, *The American Statistician*, **36**, 109.

Johnson, D. E. and F. A. Graybill (1972). An analysis of a two-way model with interaction and no replication, *Journal of the American Statistical Association*, **67**, 862.

Lauer, G. N. and C. P. Han (1974). Power of Cochran's test in the Behrens–Fisher problem, *Technometrics*, **16**, 545.

Lloyd, A. (1966). *The Making of the King*, Rinehart and Winston, New York.

Louis, T. A. (1981). Confidence intervals for a binomial parameter after observing no successes, *The American Statistician*, **35**, 154.

Mage, D. T. (1982). An objective graphical method for testing normal distributional assumptions using probability plots, *The American Statistician*, **36**, 116.

Mallows, C. L. (1973). Some comments on C_p, *Technometrics*, **15**, 661.

Martinez, J. and B. Iglewicz (1981). A test for departure from normality based on a biweight estimator of scale, *Biometrika*, **68**, 331.

McGuire, J. U., R. P. Lehmann, and A. L. Heath (1967). *Tables of Exact Probabilities for 2×2 Contingency Tests*, ARS 20-15, United States Department of Agriculture, Washington, DC.

Miller, R. G. (1977). Developments in multiple comparisons, 1966–1977, *Journal of the American Statistical Association*, **72**, 779.

Miller, R. G. (1981). *Simultaneous Statistical Inference*, 2nd ed., Springer-Verlag, New York.

Ponder, W. W. (1982). *Investigations of Model Validity Using Residuals*, Ph.D. Thesis, Department of Statistics, Iowa State University, Ames, IA.

Ramsay, J. A. (1953). Exchanges of sodium and potassium in mosquito larvae, *Journal of Experimental Biology*, **30**, 79.

Roscoe, J. T. and J. A. Byars (1971). An investigation of the restraints with respect to sample size commonly imposed on the use of the chi-square statistic, *Journal of the American Statistical Association*, **66**, 755.

Satterthwaite, F. E. (1946). An approximate distribution of estimates of variance components, *Biometrics*, **2**, 110.

Searle, S. R. (1982). *Matrix Algebra Useful for Statistics*, Wiley, New York.

Snedecor, G. W. and W. G. Cochran (1980). *Statistical Methods*, 7th ed., Iowa State University Press, Ames, IA.

Stoline, M. R. (1981). The status of multiple comparisons: Simultaneous estimation of all pairwise comparisons in one-way ANOVA, *The American Statistician*, **35**, 134.

Tate, R. F. and G. W. Klett (1959). Optimal confidence intervals for the variance of a normal distribution, *Journal of the American Statistical Association*, **54**, 674.

Tukey, J. W. (1949). One degree of freedom for non-additivity, *Biometrics*, **5**, 232.

Tukey, J. W. (1977). *Exploratory Data Analysis*, Addison-Wesley, Reading, MA.

Wetherill, G. B. (1982). *Elementary Statistical Methods*, 3rd ed., Chapman and Hall, London.

Yarnold, J. K. (1970). The minimum expectation in χ^2 goodness of fit tests and the accuracy of approximations for the null distribution, *Journal of the American Statistical Association*, **65**, 864.

Yates, F. (1934). Contingency tables involving small numbers and the χ^2-test, *Journal of the Royal Statistical Society, Supplement*, **1**, 217.

Appendix: Statistical Tables

Table A1. 2500 Computer generated random digits: The steps for making random selections of n experimental units from a collection of N units each uniquely labeled by a different one of the numbers $01, 02, \ldots, N < 100$ are:

(1) Starting point: Haphazardly choose one of the 500 groups of 5 digits, *abcde* say. With $00 \equiv 100$, take the smaller of ab and $(ab - 50)$ for the row in Table A1 and the smaller of cd and $(cd - 50)$ for the column, in which to find the starting point. Thus, $abcde = 09821$ gives the starting point as the number, 4, in row 9 and column $32 = 82 - 50$. Then:

(2) To obtain one sample of **n,** *from a total of* **N,** *units:* Suppose that $100 = kN + r$, where k is an integer and $r < N$ is the remainder. Thus, $k = 1$ if $N > 50$, $k = 4$ if $N = 24$. Then, ignoring spaces between columns and the starting number, proceed in any direction—except one steered toward convenient numbers! and regard the next two contiguous digits as one of $01, \ldots, 100$. Ignore this number if it exceeds kN, to avoid unequal frequencies: otherwise, accept the number N if the two-digit number is exactly divisible by N; if it is not, accept the remainder. Similarly, treat the next two contiguous digits and continue until n different numbers are accepted. The n correspondingly labeled experimental units comprise the sample. Thus, if a sample of $n = 5$ is required from $N = 28$ units, $k = 3$ and the sequence of paired individual contiguous digits, 89, 41, 33, 25, 61, 79, 99, 10, would generate the sample as the units with labels 13, 5, 25, 23, 10, because 89 and 99 exceed $3(28) = 84$ and the pair $61 = 2(28) + 5$ gives a previously accepted number.

(3) Random permutations: If it is required to obtain a random per-

mutation or ordering of a symbols, $A_1, A_2, \ldots, A_a$, say, the above procedure can be used with $a = n = N$. Thus, for $a = 6$, $100 = 16(6) + 4$ so that the two-digit pairs $97, \ldots, 00 = 100$ are ignored. The sequence 89, 41, 33, 25, 61, 79, 99, 10, 60 gives, after excluding 99 and dividing by 6, the remainders 5, 5, 3, 1, 1, 5, 4, $0 = 6$ and hence the permutation $A_5, A_3, A_1, A_4, A_6, A_2$.

(4) Completely randomized experiments: For a treatments with n experimental units on each, procedures (1) and (2) above give the units receiving the first treatment and repetitions, with a different starting points giving the sets of units for the remaining treatments.

(5) Randomized block designs: Symbolize the a treatments as $A_1, A_2, \ldots, A_a$ and randomly permute these using (3) above. Then the first experimental unit in the first block receives the first treatment in the permutation, the second unit in the block receives the second treatment in the permutation, and so on. Thus, if $a = 6$, the permutation in the above example would allocate treatments $A_5, A_3, A_1, A_4, A_6, A_2$ to the 1st, 2nd, $\ldots$, 6th units, respectively, in block 1. The procedure is repeated, with a different permutation, for each block.

(6) Latin square designs: First arrange the a treatment symbols, $A_1, \ldots, A_a$, in a Latin Square with each symbol once in each row and column. Then use procedure (3) to rearrange the rows and the columns in random orders and to allocate the symbols at random to the actual treatments.

Table A1. 2500 Random Digits

	1–5	6–10	11–15	16–20	21–25	26–30	31–35	36–40	41–45	46–50
1	38742	24201	25580	18631	30563	11548	08022	62261	74563	54597
2	01448	28091	45285	81470	09829	49377	88809	59780	46891	29447
3	34768	23715	37836	17206	26527	21554	62118	78918	30845	78748
4	89533	67552	74970	68065	50599	85529	20588	59726	84051	44388
5	74163	13487	64602	07271	03530	88954	66174	68319	25323	05476
6	92837	06594	01664	43011	27981	81256	75467	28245	29149	70357
7	69008	55983	22496	55337	74159	11283	13316	27479	63079	34060
8	92404	00156	38141	06269	51599	11371	24120	88150	99649	54740
9	45369	68854	67952	06245	32056	67900	84670	50098	29179	47904
10	16929	17418	70611	53752	39997	53621	67393	24891	53738	77251
11	95400	57951	64492	52389	86037	52586	42206	74681	82599	24606
12	36981	75140	26771	67681	54042	26121	70479	50295	43593	08220
13	37705	05124	60924	24374	99850	12414	13982	83219	26396	93876
14	67830	54660	89150	92919	90913	49560	49845	98239	78807	87479
15	32789	25115	44030	86301	61900	17173	34870	37043	40625	17954
16	60127	17491	59011	37625	03435	77178	08520	49910	34898	34345
17	17115	42174	81592	04300	68875	30353	48630	86132	55173	05788
18	27760	36661	85617	06242	09725	10642	44142	29625	49415	98360
19	04494	95805	16053	37126	54750	12617	09310	94021	38471	57427
20	34753	89545	33847	78318	41551	18705	64107	18200	56834	74584
21	63319	12471	56242	06344	94606	89207	26550	93261	17931	79259
22	98802	54600	92170	51425	74130	10301	08763	56046	00093	03793
23	82661	67501	01368	91079	54810	68160	11860	84288	27053	00917
24	99251	10088	48345	72786	81066	54353	17546	31595	77246	40514
25	72756	52088	29291	46169	14636	26380	35201	07490	28845	02341
26	96723	05193	38941	33288	13923	46860	12385	94973	43259	85010
27	96169	16158	24345	78561	46611	66869	17678	38209	24023	56259
28	96678	41518	88402	17882	79991	00083	29337	39994	06328	06476
29	97329	58496	55229	90839	93840	67032	77411	57137	06172	11036
30	38143	94319	58015	71878	42332	28120	80481	41745	68085	88776
31	83510	94405	93811	02145	74541	29582	24535	21485	54519	93320
32	98898	39140	50371	20646	07782	63276	66375	88305	77405	74749
33	04406	76609	46544	55985	72507	98678	48840	16601	44598	50487
34	55997	34203	29784	12914	37942	86041	48431	11784	28492	28049
35	95911	19810	65733	05412	18498	79393	37322	75911	92047	61599
36	67151	13303	12466	08918	27140	22886	61210	67131	52278	95929
37	59368	23548	60681	09171	18170	62627	48209	62135	44727	12937
38	75670	78997	76059	83474	15744	71892	52740	22930	92624	93036
39	94444	45866	42304	85506	26762	24841	47226	34746	90302	70785
40	73516	82157	24805	75928	02150	84557	12930	63123	11922	76960
41	89059	45446	56541	62549	21737	78963	30917	37046	81184	83397
42	94958	71785	47469	29362	91492	80902	80586	66162	74551	87221
43	21739	80710	61346	04257	09821	17188	80855	76589	36971	41982
44	93859	78783	46343	03715	12473	48553	02762	45114	75502	42382
45	14263	52552	17964	20078	82454	35167	35631	81815	18879	93676
46	22894	01894	47934	54594	43739	51301	22511	39456	51031	58121
47	29316	85620	09294	67074	77403	82789	22212	52358	69310	57604
48	31889	40095	98007	15605	93206	86857	29784	63937	83545	50407
49	60096	11744	74086	65948	37934	35941	25731	30787	68848	14320
50	42450	70020	43245	05233	21149	85898	73527	55648	65388	55211

Adapted from Dowdy and Wearden (1983).

Table A2. Standard, Normal, z-Variate Probabilities

$$p = P[\infty > z > z_p]$$

z_p	.00	.01	.02	.03	.04	.05	.06	.07	.08	.09
.00	.500	.496	.492	.488	.484	.480	.476	.472	.468	.464
.10	.460	.456	.452	.448	.444	.440	.436	.433	.429	.425
.20	.421	.417	.413	.409	.405	.401	.397	.394	.390	.386
.30	.382	.378	.374	.371	.367	.363	.359	.356	.352	.348
.40	.345	.341	.337	.334	.330	.326	.323	.319	.316	.312
.50	.309	.305	.302	.298	.295	.291	.288	.284	.281	.278
.60	.274	.271	.268	.264	.261	.258	.255	.251	.248	.245
.70	.242	.239	.236	.233	.230	.227	.224	.221	.218	.215
.80	.212	.209	.206	.203	.200	.198	.195	.192	.189	.187
.90	.184	.181	.179	.176	.174	.171	.169	.166	.164	.161
1.00	.159	.156	.154	.152	.149	.147	.145	.142	.140	.138
1.10	.136	.133	.131	.129	.127	.125	.123	.121	.119	.117
1.20	.115	.113	.111	.109	.107	.106	.104	.102	.100	.099
1.30	.097	.095	.093	.092	.090	.089	.087	.085	.084	.082
1.40	.081	.079	.078	.076	.075	.074	.072	.071	.069	.068
1.50	.067	.066	.064	.063	.062	.061	.059	.058	.057	.056
1.60	.055	.054	.053	.052	.051	.049	.048	.047	.046	.046
1.70	.045	.044	.043	.042	.041	.040	.039	.038	.038	.037
1.80	.036	.035	.034	.034	.033	.032	.031	.031	.030	.029
1.90	.029	.028	.027	.027	.026	.026	.025	.024	.024	.023
2.00	.023	.022	.022	.021	.021	.020	.020	.019	.019	.018
2.10	.018	.017	.017	.017	.016	.016	.015	.015	.015	.014
2.20	.014	.014	.013	.013	.013	.012	.012	.012	.011	.011
2.30	.011	.010	.010	.010	.010	.009	.009	.009	.009	.008
2.40	.008	.008	.008	.008	.007	.007	.007	.007	.007	.006
2.50	.006	.006	.006	.006	.006	.005	.005	.005	.005	.005
2.60	.005	.005	.004	.004	.004	.004	.004	.004	.004	.004
2.70	.003	.003	.003	.003	.003	.003	.003	.003	.003	.003
2.80	.003	.002	.002	.002	.002	.002	.002	.002	.002	.002
2.90	.002	.002	.002	.002	.002	.002	.002	.001	.001	.001
3.00	.001	.001	.001	.001	.001	.001	.001	.001	.001	.001

Adapted from Dowdy and Wearden (1983).

Table A3. *t*-Variate Probabilities

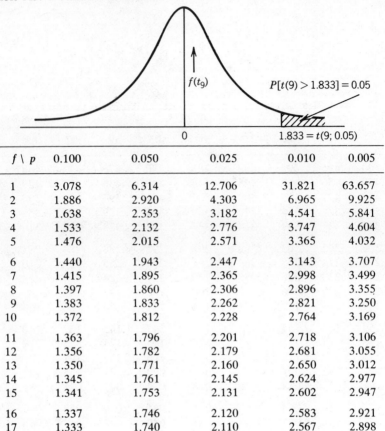

$f \setminus p$	0.100	0.050	0.025	0.010	0.005
1	3.078	6.314	12.706	31.821	63.657
2	1.886	2.920	4.303	6.965	9.925
3	1.638	2.353	3.182	4.541	5.841
4	1.533	2.132	2.776	3.747	4.604
5	1.476	2.015	2.571	3.365	4.032
6	1.440	1.943	2.447	3.143	3.707
7	1.415	1.895	2.365	2.998	3.499
8	1.397	1.860	2.306	2.896	3.355
9	1.383	1.833	2.262	2.821	3.250
10	1.372	1.812	2.228	2.764	3.169
11	1.363	1.796	2.201	2.718	3.106
12	1.356	1.782	2.179	2.681	3.055
13	1.350	1.771	2.160	2.650	3.012
14	1.345	1.761	2.145	2.624	2.977
15	1.341	1.753	2.131	2.602	2.947
16	1.337	1.746	2.120	2.583	2.921
17	1.333	1.740	2.110	2.567	2.898
18	1.330	1.734	2.101	2.552	2.878
19	1.328	1.729	2.093	2.539	2.861
20	1.325	1.725	2.086	2.528	2.845
21	1.323	1.721	2.080	2.518	2.831
22	1.321	1.717	2.074	2.508	2.819
23	1.319	1.714	2.069	2.500	2.807
24	1.318	1.711	2.064	2.492	2.797
25	1.316	1.708	2.060	2.485	2.787
26	1.315	1.706	2.056	2.479	2.779
27	1.314	1.703	2.052	2.473	2.771
28	1.313	1.701	2.048	2.467	2.763
29	1.311	1.699	2.045	2.462	2.756
30	1.310	1.697	2.042	2.457	2.750
40	1.303	1.684	2.021	2.423	2.704
60	1.296	1.671	2.000	2.390	2.660
120	1.289	1.658	1.980	2.358	2.617
∞	1.282	1.645	1.960	2.326	2.576

Adapted from Dowdy and Wearden (1983).

Table A4. χ^2-Variate Probabilities

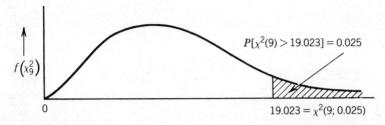

$$P[\chi^2(9) > 19.023] = 0.025$$

$$19.023 = \chi^2(9; 0.025)$$

$f \setminus p$	0.995	0.990	0.975	0.950	0.050	0.025	0.010	0.005
1	0.000	0.000	0.001	0.004	3.841	5.024	6.635	7.879
2	0.010	0.020	0.051	0.103	5.991	7.378	9.210	10.597
3	0.072	0.115	0.216	0.352	7.815	9.348	11.345	12.838
4	0.207	0.297	0.448	0.711	9.488	11.143	13.277	14.860
5	0.412	0.554	0.831	1.145	11.070	12.833	15.086	16.750
6	0.676	0.872	1.237	1.635	12.592	14.449	16.812	18.548
7	0.989	1.239	1.690	2.167	14.067	16.013	18.475	20.278
8	1.344	1.646	2.180	2.733	15.507	17.535	20.090	21.955
9	1.735	2.088	2.700	3.325	16.919	19.023	21.666	23.589
10	2.156	2.558	3.247	3.940	18.307	20.483	23.209	25.188
11	2.603	3.053	3.816	4.575	19.675	21.920	24.725	26.757
12	3.074	3.571	4.404	5.226	21.026	23.337	26.217	28.300
13	3.565	4.107	5.009	5.892	22.362	24.736	27.688	29.819
14	4.075	4.660	5.629	6.571	23.685	26.119	29.141	31.319
15	4.601	5.229	6.262	7.261	24.996	27.488	30.578	32.801
16	5.142	5.812	6.908	7.962	26.296	28.845	32.000	34.267
17	5.697	6.408	7.564	8.672	27.587	30.191	33.409	35.718
18	6.265	7.015	8.231	9.390	28.869	31.526	34.805	37.156
19	6.844	7.633	8.907	10.117	30.144	32.852	36.191	38.852
20	7.434	8.260	9.591	10.851	31.410	34.170	37.566	39.997
21	8.034	8.897	10.283	11.591	32.671	35.479	38.932	41.401
22	8.643	9.542	10.982	12.338	33.924	36.781	40.289	42.796
23	9.260	10.196	11.689	13.091	35.172	38.076	41.638	44.181
24	9.886	10.856	12.401	13.848	36.415	39.364	42.980	45.559
25	10.520	11.524	13.120	14.611	37.652	40.646	44.314	46.928
26	11.160	12.198	13.844	15.379	38.885	41.923	45.642	48.290
27	11.808	12.879	14.573	16.151	40.113	43.195	46.963	49.645
28	12.461	13.565	15.308	16.928	41.337	44.461	48.278	50.993
29	13.121	14.256	16.047	17.708	42.557	45.722	49.588	52.336
30	13.787	14.953	16.791	18.493	43.773	46.979	50.892	53.672
32	15.134	16.362	18.291	20.072	46.194	49.480	53.486	56.328
34	16.501	17.789	19.806	21.664	48.602	51.966	56.061	58.964
36	17.887	19.233	21.336	23.269	50.998	54.437	58.619	61.581
38	19.289	20.691	22.878	24.884	53.384	56.896	61.162	64.181
40	20.707	22.164	24.433	26.509	55.758	59.342	63.691	66.766
42	22.138	23.650	25.999	28.144	58.124	61.777	66.206	69.336
44	23.584	25.148	27.575	29.787	60.481	64.201	68.710	71.893
46	25.041	26.657	29.160	31.439	62.830	66.617	71.201	74.437
48	26.511	28.177	30.755	33.098	65.171	69.023	73.683	76.969
50	27.991	29.707	32.357	34.764	67.505	71.420	76.154	79.490
60	35.534	37.485	40.482	43.188	79.082	83.298	88.379	91.952
70	43.275	45.442	48.758	51.739	90.531	95.023	100.425	104.215
80	51.172	53.540	57.153	60.391	101.879	106.629	112.329	116.321
90	59.196	61.754	65.647	69.126	113.145	118.136	124.116	128.299
100	67.328	70.065	74.222	77.929	124.342	129.561	135.807	140.169

Adapted from Dowdy and Wearden (1983).

Table A5. *F*-Variate Probabilities (*p* in column 2, F_p in body)

Denomi-nator df (f_2)	Probability of a larger F	Numerator df (f_1)								
		1	2	3	4	5	6	7	8	9
1	.100	39.86	49.50	53.59	55.83	57.24	58.20	58.91	59.44	59.86
	.050	161.4	199.5	215.7	224.6	230.2	234.0	236.8	238.9	240.5
	.025	647.8	799.5	864.6	899.6	921.8	937.1	948.2	956.7	963.3
	.010	4052	4999.5	5403	5625	5764	5859	5928	5982	6022
	.005	16211	20000	21615	22500	23056	23437	23715	23925	24091
2	.100	8.53	9.00	9.16	9.24	9.29	9.33	9.35	9.37	9.38
	.050	18.51	19.00	19.16	19.25	19.30	19.33	19.35	19.37	19.38
	.025	38.51	39.00	39.17	39.25	39.30	·39.33	39.36	39.37	39.39
	.010	98.50	99.00	99.17	99.25	99.30	99.33	99.36	99.37	99.39
	.005	198.5	199.0	199.2	199.2	199.3	199.3	199.4	199.4	199.4
3	.100	5.54	5.46	5.39	5.34	5.31	5.28	5.27	5.25	5.24
	.050	10.13	9.55	9.28	9.12	9.01	8.94	8.89	8.85	8.81
	.025	17.44	16.04	15.44	15.10	14.88	14.73	14.62	14.54	14.47
	.010	34.12	30.82	29.46	28.71	28.24	27.91	27.67	27.49	27.35
	.005	55.55	49.80	47.47	46.19	45.39	44.84	44.43	44.13	43.88
4	.100	4.54	4.32	4.19	4.11	4.05	4.01	3.98	3.95	3.94
	.050	7.71	6.94	6.59	6.39	6.26	6.16	6.09	6.04	6.00
	.025	12.22	10.65	9.98	9.60	9.36	9.20	9.07	8.98	8.90
	.010	21.20	18.00	16.69	15.98	15.52	15.21	14.98	14.80	14.66
	.005	31.33	26.28	24.26	23.15	22.46	21.97	21.62	21.35	21.14
5	.100	4.06	3.78	3.62	3.52	3.45	3.40	3.37	3.34	3.32
	.050	6.61	5.79	5.41	5.19	5.05	4.95	4.88	4.82	4.77
	.025	10.01	8.43	7.76	7.39	7.15	6.98	6.85	6.76	6.68
	.010	16.26	13.27	12.06	11.39	10.97	10.67	10.46	10.29	10.16
	.005	22.78	18.31	16.53	15.56	14.94	14.51	14.20	13.96	13.77
6	.100	3.78	3.46	3.29	3.18	3.11	3.05	3.01	2.98	2.96
	.050	5.99	5.14	4.76	4.53	4.39	4.28	4.21	4.15	4.10
	.025	8.81	7.26	6.60	6.23	5.99	5.82	5.70	5.60	5.52
	.010	13.75	10.92	9.78	9.15	8.75	8.47	8.26	8.10	7.98
	.005	18.63	14.54	12.92	12.03	11.46	11.07	10.79	10.57	10.39
7	.100	3.59	3.26	3.07	2.96	2.88	2.83	2.78	2.75	2.72
	.050	5.59	4.74	4.35	4.12	3.97	3.87	3.79	3.73	3.68
	.025	8.07	6.54	5.89	5.52	5.29	5.12	4.99	4.90	4.82
	.010	12.25	9.55	8.45	7.85	7.46	7.19	6.99	6.84	6.72
	.005	16.24	12.40	10.88	10.05	9.52	9.16	8.89	8.68	8.51
8	.100	3.46	3.11	2.92	2.81	2.73	2.67	2.62	2.59	2.56
	.050	5.32	4.46	4.07	3.84	3.69	3.58	3.50	3.44	3.39
	.025	7.57	6.06	5.42	5.05	4.82	4.65	4.53	4.43	4.36
	.010	11.26	8.65	7.59	7.01	6.63	6.37	6.18	6.03	5.91
	.005	14.69	11.04	9.60	8.81	8.30	7.95	7.69	7.50	7.34
9	.100	3.36	3.01	2.81	2.69	2.61	2.55	2.51	2.47	2.44
	.050	5.12	4.26	3.86	3.63	3.48	3.37	3.29	3.23	3.18
	.025	7.21	5.71	5.08	4.72	4.48	4.32	4.20	4.10	4.03
	.010	10.56	8.02	6.99	6.42	6.06	5.80	5.61	5.47	5.35
	.005	13.61	10.11	8.72	7.96	7.47	7.13	6.88	6.69	6.54
10	.100	3.29	2.92	2.73	2.61	2.52	2.46	2.41	2.38	2.35
	.050	4.96	4.10	3.71	3.48	3.33	3.22	3.14	3.07	3.02
	.025	6.94	5.46	4.83	4.47	4.24	4.07	3.95	3.85	3.78
	.010	10.04	7.56	6.55	5.99	5.64	5.39	5.20	5.06	4.94
	.005	12.83	9.43	8.08	7.34	6.87	6.54	6.30	6.12	5.97
11	.100	3.23	2.86	2.66	2.54	2.45	2.39	2.34	2.30	2.27
	.050	4.84	3.98	3.59	3.36	3.20	3.09	3.01	2.95	2.90
	.025	6.72	5.26	4.63	4.28	4.04	3.88	3.76	3.66	3.59
	.010	9.65	7.21	6.22	5.67	5.32	5.07	4.89	4.74	4.63
	.005	12.23	8.91	7.60	6.88	6.42	6.10	5.86	5.68	5.54
12	.100	3.18	2.81	2.61	2.48	2.39	2.33	2.28	2.24	2.21
	.050	4.75	3.89	3.49	3.26	3.11	3.00	2.91	2.85	2.80
	.025	6.55	5.10	4.47	4.12	3.89	3.73	3.61	3.51	3.44
	.010	9.33	6.93	5.95	5.41	5.06	4.82	4.64	4.50	4.39
	.005	11.75	8.51	7.23	6.52	6.07	5.76	5.52	5.35	5.20
13	.100	3.14	2.76	2.56	2.43	2.35	2.28	2.23	2.20	2.16
	.050	4.67	3.81	3.41	3.18	3.03	2.92	2.83	2.77	2.71
	.025	6.41	4.97	4.35	4.00	3.77	3.60	3.48	3.39	3.31
	.010	9.07	6.70	5.74	5.21	4.86	4.62	4.44	4.30	4.19
	.005	11.37	8.19	6.93	6.23	5.79	5.48	5.25	5.08	4.94
14	.100	3.10	2.73	2.52	2.39	2.31	2.24	2.19	2.15	2.12
	.050	4.60	3.74	3.34	3.11	2.96	2.85	2.76	2.70	2.65
	.025	6.30	4.86	4.24	3.89	3.66	3.50	3.38	3.29	3.21
	.010	8.86	6.51	5.56	5.04	4.69	4.46	4.28	4.14	4.03
	.005	11.06	7.92	6.68	6.00	5.56	5.26	5.03	4.86	4.72

Table A5. (*Continued*)

				Numerator df (f_1)							
10	12	15	20	24	30	40	60	120	∞	*p*	f_2
60.19	60.71	61.22	61.74	62.00	62.26	62.53	62.79	63.06	63.33	.100	1
241.9	243.9	245.9	248.0	249.1	250.1	251.1	252.2	253.3	254.3	.050	
968.6	976.7	984.9	993.1	997.2	1001	1006	1010	1014	1018	.025	
6056	6106	6157	6209	6235	6261	6287	6313	6339	6366	.010	
24224	24426	24630	24836	24940	25044	25148	25253	25359	25465	.005	
9.39	9.41	9.42	9.44	9.45	9.46	9.47	9.47	9.48	9.49	.100	2
19.40	19.41	19.43	19.45	19.45	19.46	19.47	19.48	19.49	19.50	.050	
39.40	39.41	39.43	39.45	39.46	39.46	39.47	39.48	39.49	39.50	.025	
99.40	99.42	99.43	99.45	99.46	99.47	99.47	99.48	99.49	99.50	.010	
199.4	199.4	199.4	199.4	199.5	199.5	199.5	199.5	199.5	199.5	.005	
5.23	5.22	5.20	5.18	5.18	5.17	5.16	5.15	5.14	5.13	.100	3
8.79	8.74	8.70	8.66	8.64	8.62	8.59	8.57	8.55	8.53	.050	
14.42	14.34	14.25	14.17	14.12	14.08	14.04	13.99	13.95	13.90	.025	
27.23	27.05	26.87	26.69	26.60	26.50	26.41	26.32	26.22	26.13	.010	
43.69	43.39	43.08	42.78	42.62	42.47	42.31	42.15	41.99	41.83	.005	
3.92	3.90	3.87	3.84	3.83	3.82	3.80	3.79	3.78	3.76	.100	4
5.96	5.91	5.86	5.80	5.77	5.75	5.72	5.69	5.66	5.63	.050	
8.84	8.75	8.66	8.56	8.51	8.46	8.41	8.36	8.31	8.26	.025	
14.55	14.37	14.20	14.02	13.93	13.84	13.75	13.65	13.56	13.46	.010	
20.97	20.70	20.44	20.17	20.03	19.89	19.75	19.61	19.47	19.32	.005	
3.30	3.27	3.24	3.21	3.19	3.17	3.16	3.14	3.12	3.10	.100	5
4.74	4.68	4.62	4.56	4.53	4.50	4.46	4.43	4.40	4.36	.050	
6.62	6.52	6.43	6.33	6.28	6.23	6.18	6.12	6.07	6.02	.025	
10.05	9.89	9.72	9.55	9.47	9.38	9.29	9.20	9.11	9.02	.010	
13.62	13.38	13.15	12.90	12.78	12.66	12.53	12.40	12.27	12.14	.005	
2.94	2.90	2.87	2.84	2.82	2.80	2.78	2.76	2.74	2.72	.100	6
4.06	4.00	3.94	3.87	3.84	3.81	3.77	3.74	3.70	3.67	.050	
5.46	5.37	5.27	5.17	5.12	5.07	5.01	4.96	4.90	4.85	.025	
7.87	7.72	7.56	7.40	7.31	7.23	7.14	7.06	6.97	6.88	.010	
10.25	10.03	9.81	9.59	9.47	9.36	9.24	9.12	9.00	8.88	.005	
2.70	2.67	2.63	2.59	2.58	2.56	2.54	2.51	2.49	2.47	.100	7
3.64	3.57	3.51	3.44	3.41	3.38	3.34	3.30	3.27	3.23	.050	
4.76	4.67	4.57	4.47	4.42	4.36	4.31	4.25	4.20	4.14	.025	
6.62	6.47	6.31	6.16	6.07	5.99	5.91	5.82	5.74	5.65	.010	
8.38	8.18	7.97	7.75	7.65	7.53	7.42	7.31	7.19	7.08	.005	
2.54	2.50	2.46	2.42	2.40	2.38	2.36	2.34	2.32	2.29	.100	8
3.35	3.28	3.22	3.15	3.12	3.08	3.04	3.01	2.97	2.93	.050	
4.30	4.20	4.10	4.00	3.95	3.89	3.84	3.78	3.73	3.67	.025	
5.81	5.67	5.52	5.36	5.28	5.20	5.12	5.03	4.95	4.86	.010	
7.21	7.01	6.81	6.61	6.50	6.40	6.29	6.18	6.06	5.95	.005	
2.42	2.38	2.34	2.30	2.28	2.25	2.23	2.21	2.18	2.16	.100	9
3.14	3.07	3.01	2.94	2.90	2.86	2.83	2.79	2.75	2.71	.050	
3.96	3.87	3.77	3.67	3.61	3.56	3.51	3.45	3.39	3.33	.025	
5.26	5.11	4.96	4.81	4.73	4.65	4.57	4.48	4.40	4.31	.010	
6.42	6.23	6.03	5.83	5.73	5.62	5.52	5.41	5.30	5.19	.005	
2.32	2.28	2.24	2.20	2.18	2.16	2.13	2.11	2.08	2.06	.100	10
2.98	2.91	2.85	2.77	2.74	2.70	2.66	2.62	2.58	2.54	.050	
3.72	3.62	3.52	3.42	3.37	3.31	3.26	3.20	3.14	3.08	.025	
4.85	4.71	4.56	4.41	4.33	4.25	4.17	4.08	4.00	3.91	.010	
5.85	5.66	5.47	5.27	5.17	5.07	4.97	4.86	4.75	4.64	.005	
2.25	2.21	2.17	2.12	2.10	2.08	2.05	2.03	2.00	1.97	.100	11
2.85	2.79	2.72	2.65	2.61	2.57	2.53	2.49	2.45	2.40	.050	
3.53	3.43	3.33	3.23	3.17	3.12	3.06	3.00	2.94	2.88	.025	
4.54	4.40	4.25	4.10	4.02	3.94	3.86	3.78	3.69	3.60	.010	
5.42	5.24	5.05	4.86	4.76	4.65	4.55	4.44	4.34	4.23	.005	
2.19	2.15	2.10	2.06	2.04	2.01	1.99	1.96	1.93	1.90	.100	12
2.75	2.69	2.62	2.54	2.51	2.47	2.43	2.38	2.34	2.30	.050	
3.37	3.28	3.18	3.07	3.02	2.96	2.91	2.85	2.79	2.72	.025	
4.30	4.16	4.01	3.86	3.78	3.70	3.62	3.54	3.45	3.36	.010	
5.09	4.91	4.72	4.53	4.43	4.33	4.23	4.12	4.01	3.90	.005	
2.14	2.10	2.05	2.01	1.98	1.96	1.93	1.90	1.88	1.85	.100	13
2.67	2.60	2.53	2.46	2.42	2.38	2.34	2.30	2.25	2.21	.050	
3.25	3.15	3.05	2.95	2.89	2.84	2.78	2.72	2.66	2.60	.025	
4.10	3.96	3.82	3.66	3.59	3.51	3.43	3.34	3.25	3.17	.010	
4.82	4.64	4.46	4.27	4.17	4.07	3.97	3.87	3.76	3.65	.005	
2.10	2.05	2.01	1.96	1.94	1.91	1.89	1.86	1.83	1.80	.100	14
2.60	2.53	2.46	2.39	2.35	2.31	2.27	2.22	2.18	2.13	.050	
3.15	3.05	2.95	2.84	2.79	2.73	2.67	2.61	2.55	2.49	.025	
3.94	3.80	3.66	3.51	3.43	3.35	3.27	3.18	3.09	3.00	.010	
4.60	4.43	4.25	4.06	3.96	3.86	3.76	3.66	3.55	3.44	.005	

Table A5. (*Continued*)

Denomi-nator df (f_2)	Probability of a Larger F	Numerator df (f_1)								
		1	2	3	4	5	6	7	8	9
15	.100	3.07	2.70	2.49	2.36	2.27	2.21	2.16	2.12	2.09
	.050	4.54	3.68	3.29	3.06	2.90	2.79	2.71	2.64	2.59
	.025	6.20	4.77	4.15	3.80	3.58	3.41	3.29	3.20	3.12
	.010	8.68	6.36	5.42	4.89	4.56	4.32	4.14	4.00	3.89
	.005	10.80	7.70	6.48	5.80	5.37	5.07	4.85	4.67	4.54
16	.100	3.05	2.67	2.46	2.33	2.24	2.18	2.13	2.09	2.06
	.050	4.49	3.63	3.24	3.01	2.85	2.74	2.66	2.59	2.54
	.025	6.12	4.69	4.08	3.73	3.50	3.34	3.22	3.12	3.05
	.010	8.53	6.23	5.29	4.77	4.44	4.20	4.03	3.89	3.78
	.005	10.58	7.51	6.30	5.64	5.21	4.91	4.69	4.52	4.38
17	.100	3.03	2.64	2.44	2.31	2.22	2.15	2.10	2.06	2.03
	.050	4.45	3.59	3.20	2.96	2.81	2.70	2.61	2.55	2.49
	.025	6.04	4.62	4.01	3.66	3.44	3.28	3.16	3.06	2.98
	.010	8.40	6.11	5.18	4.67	4.34	4.10	3.93	3.79	3.68
	.005	10.38	7.35	6.16	5.50	5.07	4.78	4.56	4.39	4.25
18	.100	3.01	2.62	2.42	2.29	2.20	2.13	2.08	2.04	2.00
	.050	4.41	3.55	3.16	2.93	2.77	2.66	2.58	2.51	2.46
	.025	5.98	4.56	3.95	3.61	3.38	3.22	3.10	3.01	2.93
	.010	8.29	6.01	5.09	4.58	4.25	4.01	3.84	3.71	3.60
	.005	10.22	7.21	6.03	5.37	4.96	4.66	4.44	4.28	4.14
19	.100	2.99	2.61	2.40	2.27	2.18	2.11	2.06	2.02	1.98
	.050	4.38	3.52	3.13	2.90	2.74	2.63	2.54	2.48	2.42
	.025	5.92	4.51	3.90	3.56	3.33	3.17	3.05	2.96	2.88
	.010	8.18	5.93	5.01	4.50	4.17	3.94	3.77	3.63	3.52
	.005	10.07	7.09	5.92	5.27	4.85	4.56	4.34	4.18	4.04
20	.100	2.97	2.59	2.38	2.25	2.16	2.09	2.04	2.00	1.96
	.050	4.35	3.49	3.10	2.87	2.71	2.60	2.51	2.45	2.39
	.025	5.87	4.46	3.86	3.51	3.29	3.13	3.01	2.91	2.84
	.010	8.10	5.85	4.94	4.43	4.10	3.87	3.70	3.56	3.46
	.005	9.94	6.99	5.82	5.17	4.76	4.47	4.26	4.09	3.96
21	.100	2.96	2.57	2.36	2.23	2.14	2.08	2.02	1.98	1.95
	.050	4.32	3.47	3.07	2.84	2.68	2.57	2.49	2.42	2.37
	.025	5.83	4.42	3.82	3.48	3.25	3.09	2.97	2.87	2.80
	.010	8.02	5.78	4.87	4.37	4.04	3.81	3.64	3.51	3.40
	.005	9.83	6.89	5.73	5.09	4.68	4.39	4.18	4.01	3.88
22	.100	2.95	2.56	2.35	2.22	2.13	2.06	2.01	1.97	1.93
	.050	4.30	3.44	3.05	2.82	2.66	2.55	2.46	2.40	2.34
	.025	5.79	4.38	3.78	3.44	3.22	3.05	2.93	2.84	2.76
	.010	7.95	5.72	4.82	4.31	3.99	3.76	3.59	3.45	3.35
	.005	9.73	6.81	5.65	5.02	4.61	4.32	4.11	3.94	3.81
23	.100	2.94	2.55	2.34	2.21	2.11	2.05	1.99	1.95	1.92
	.050	4.28	3.42	3.03	2.80	2.64	2.53	2.44	2.37	2.32
	.025	5.75	4.35	3.75	3.41	3.18	3.02	2.90	2.81	2.73
	.010	7.88	5.66	4.76	4.26	3.94	3.71	3.54	3.41	3.30
	.005	9.63	6.73	5.58	4.95	4.54	4.26	4.05	3.88	3.75
24	.100	2.93	2.54	2.33	2.19	2.10	2.04	1.98	1.94	1.91
	.050	4.26	3.40	3.01	2.78	2.62	2.51	2.42	2.36	2.30
	.025	5.72	4.32	3.72	3.38	3.15	2.99	2.87	2.78	2.70
	.010	7.82	5.61	4.72	4.22	3.90	3.67	3.50	3.36	3.26
	.005	9.55	6.66	5.52	4.89	4.49	4.20	3.99	3.83	3.69
25	.100	2.92	2.53	2.32	2.18	2.09	2.02	1.97	1.93	1.89
	.050	4.24	3.39	2.99	2.76	2.60	2.49	2.40	2.34	2.28
	.025	5.69	4.29	3.69	3.35	3.13	2.97	2.85	2.75	2.68
	.010	7.77	5.57	4.68	4.18	3.85	3.63	3.46	3.32	3.22
	.005	9.48	6.60	5.46	4.84	4.43	4.15	3.94	3.78	3.64
26	.100	2.91	2.52	2.31	2.17	2.08	2.01	1.96	1.92	1.88
	.050	4.23	3.37	2.98	2.74	2.59	2.47	2.39	2.32	2.27
	.025	5.66	4.27	3.67	3.33	3.10	2.94	2.82	2.73	2.65
	.010	7.72	5.53	4.64	4.14	3.82	3.59	3.42	3.29	3.18
	.005	9.41	6.54	5.41	4.79	4.38	4.10	3.89	3.73	3.60
27	.100	2.90	2.51	2.30	2.17	2.07	2.00	1.95	1.91	1.87
	.050	4.21	3.35	2.96	2.73	2.57	2.46	2.37	2.31	2.25
	.025	5.63	4.24	3.65	3.31	3.08	2.92	2.80	2.71	2.63
	.010	7.68	5.49	4.60	4.11	3.78	3.56	3.39	3.26	3.15
	.005	9.34	6.49	5.36	4.74	4.34	4.06	3.85	3.69	3.56
28	.100	2.89	2.50	2.29	2.16	2.06	2.00	1.94	1.90	1.87
	.050	4.20	3.34	2.95	2.71	2.56	2.45	2.36	2.29	2.24
	.025	5.61	4.22	3.63	3.29	3.06	2.90	2.78	2.69	2.61
	.010	7.64	5.45	4.57	4.07	3.75	3.53	3.36	3.23	3.12
	.005	9.28	6.44	5.32	4.70	4.30	4.02	3.81	3.65	3.52

Table A5. (*Continued*)

				Numerator df (f_1)							
10	12	15	20	24	30	40	60	120	∞	p	f_2
2.06	2.02	1.97	1.92	1.90	1.87	1.85	1.82	1.79	1.76	.100	15
2.54	2.48	2.40	2.33	2.29	2.25	2.20	2.16	2.11	2.07	.050	
3.06	2.96	2.86	2.76	2.70	2.64	2.59	2.52	2.46	2.40	.025	
3.80	3.67	3.52	3.37	3.29	3.21	3.13	3.05	2.96	2.87	.010	
4.42	4.25	4.07	3.88	3.79	3.69	3.58	3.48	3.37	3.26	.005	
2.03	1.99	1.94	1.89	1.87	1.84	1.81	1.78	1.75	1.72	.100	16
2.49	2.42	2.35	2.28	2.24	2.19	2.15	2.11	2.06	2.01	.050	
2.99	2.89	2.79	2.68	2.63	2.57	2.51	2.45	2.38	2.32	.025	
3.69	3.55	3.41	3.26	3.18	3.10	3.02	2.93	2.84	2.75	.010	
4.27	4.10	3.92	3.73	3.64	3.54	3.44	3.33	3.22	3.11	.005	
2.00	1.96	1.91	1.86	1.84	1.81	1.78	1.75	1.72	1.69	.100	17
2.45	2.38	2.31	2.23	2.19	2.15	2.10	2.06	2.01	1.96	.050	
2.92	2.82	2.72	2.62	2.56	2.50	2.44	2.38	2.32	2.25	.025	
3.59	3.46	3.31	3.16	3.08	3.00	2.92	2.83	2.75	2.65	.010	
4.14	3.97	3.79	3.61	3.51	3.41	3.31	3.21	3.10	2.98	.005	
1.98	1.93	1.89	1.84	1.81	1.78	1.75	1.72	1.69	1.66	.100	18
2.41	2.34	2.27	2.19	2.15	2.11	2.06	2.02	1.97	1.92	.050	
2.87	2.77	2.67	2.56	2.50	2.44	2.38	2.32	2.26	2.19	.025	
3.51	3.37	3.23	3.08	3.00	2.92	2.84	2.75	2.66	2.57	.010	
4.03	3.86	3.68	3.50	3.40	3.30	3.20	3.10	2.99	2.87	.005	
1.96	1.91	1.86	1.81	1.79	1.76	1.73	1.70	1.67	1.63	.100	19
2.38	2.31	2.23	2.16	2.11	2.07	2.03	1.98	1.93	1.88	.050	
2.82	2.72	2.62	2.51	2.45	2.39	2.33	2.27	2.20	2.13	.025	
3.43	3.30	3.15	3.00	2.92	2.84	2.76	2.67	2.58	2.49	.010	
3.93	3.76	3.59	3.40	3.31	3.21	3.11	3.00	2.89	2.78	.005	
1.94	1.89	1.84	1.79	1.77	1.74	1.71	1.68	1.64	1.61	.100	20
2.35	2.28	2.20	2.12	2.08	2.04	1.99	1.95	1.90	1.84	.050	
2.77	2.68	2.57	2.46	2.41	2.35	2.29	2.22	2.16	2.09	.025	
3.37	3.23	3.09	2.94	2.86	2.78	2.69	2.61	2.52	2.42	.010	
3.85	3.68	3.50	3.32	3.22	3.12	3.02	2.92	2.81	2.69	.005	
1.92	1.87	1.83	1.78	1.75	1.72	1.69	1.66	1.62	1.59	.100	21
2.32	2.25	2.18	2.10	2.05	2.01	1.96	1.92	1.87	1.81	.050	
2.73	2.64	2.53	2.42	2.37	2.31	2.25	2.18	2.11	2.04	.025	
3.31	3.17	3.03	2.88	2.80	2.72	2.64	2.55	2.46	2.36	.010	
3.77	3.60	3.43	3.24	3.15	3.05	2.95	2.84	2.73	2.61	.005	
1.90	1.86	1.81	1.76	1.73	1.70	1.67	1.64	1.60	1.57	.100	22
2.30	2.23	2.15	2.07	2.03	1.98	1.94	1.89	1.84	1.78	.050	
2.70	2.60	2.50	2.39	2.33	2.27	2.21	2.14	2.08	2.00	.025	
3.26	3.12	2.98	2.83	2.75	2.67	2.58	2.50	2.40	2.31	.010	
3.70	3.54	3.36	3.18	3.08	2.98	2.88	2.77	2.66	2.55	.005	
1.89	1.84	1.80	1.74	1.72	1.69	1.66	1.62	1.59	1.55	.100	23
2.27	2.20	2.13	2.05	2.01	1.96	1.91	1.86	1.81	1.76	.050	
2.67	2.57	2.47	2.36	2.30	2.24	2.18	2.11	2.04	1.97	.025	
3.21	3.07	2.93	2.78	2.70	2.62	2.54	2.45	2.35	2.26	.010	
3.64	3.47	3.30	3.12	3.02	2.92	2.82	2.71	2.60	2.48	.005	
1.88	1.83	1.78	1.73	1.70	1.67	1.64	1.61	1.57	1.53	.100	24
2.25	2.18	2.11	2.03	1.98	1.94	1.89	1.84	1.79	1.73	.050	
2.64	2.54	2.44	2.33	2.27	2.21	2.15	2.08	2.01	1.94	.025	
3.17	3.03	2.89	2.74	2.66	2.58	2.49	2.40	2.31	2.21	.010	
3.59	3.42	3.25	3.06	2.97	2.87	2.77	2.66	2.55	2.43	.005	
1.87	1.82	1.77	1.72	1.69	1.66	1.63	1.59	1.56	1.52	.100	25
2.24	2.16	2.09	2.01	1.96	1.92	1.87	1.82	1.77	1.71	.050	
2.61	2.51	2.41	2.30	2.24	2.18	2.12	2.05	1.98	1.91	.025	
3.13	2.99	2.85	2.70	2.62	2.54	2.45	2.36	2.27	2.17	.010	
3.54	3.37	3.20	3.01	2.92	2.82	2.72	2.61	2.50	2.38	.005	
1.86	1.81	1.76	1.71	1.68	1.65	1.61	1.58	1.54	1.50	.100	26
2.22	2.15	2.07	1.99	1.95	1.90	1.85	1.80	1.75	1.69	.050	
2.59	2.49	2.39	2.28	2.22	2.16	2.09	2.03	1.95	1.88	.025	
3.09	2.96	2.81	2.66	2.58	2.50	2.42	2.33	2.23	2.13	.010	
3.49	3.33	3.15	2.97	2.87	2.77	2.67	2.56	2.45	2.33	.005	
1.85	1.80	1.75	1.70	1.67	1.64	1.60	1.57	1.53	1.49	.100	27
2.20	2.13	2.06	1.97	1.93	1.88	1.84	1.79	1.73	1.67	.050	
2.57	2.47	2.36	2.25	2.19	2.13	2.07	2.00	1.93	1.85	.025	
3.06	2.93	2.78	2.63	2.55	2.47	2.38	2.29	2.20	2.10	.010	
3.45	3.28	3.11	2.93	2.83	2.73	2.63	2.52	2.41	2.29	.005	
1.84	1.79	1.74	1.69	1.66	1.63	1.59	1.56	1.52	1.48	.100	28
2.19	2.12	2.04	1.96	1.91	1.87	1.82	1.77	1.71	1.65	.050	
2.55	2.45	2.34	2.23	2.17	2.11	2.05	1.98	1.91	1.83	.025	
3.03	2.90	2.75	2.60	2.52	2.44	2.35	2.26	2.17	2.06	.010	
3.41	3.25	3.07	2.89	2.79	2.69	2.59	2.48	2.37	2.25	.005	

Denomi-nator df (f_2)	Probability of a Larger F	Numerator df (f_1)								
		1	2	3	4	5	6	7	8	9
29	.100	2.89	2.50	2.28	2.15	2.06	1.99	1.93	1.89	1.86
	.050	4.18	3.33	2.93	2.70	2.55	2.43	2.35	2.28	2.22
	.025	5.59	4.20	3.61	3.27	3.04	2.88	2.76	2.67	2.59
	.010	7.60	5.42	4.54	4.04	3.73	3.50	3.33	3.20	3.09
	.005	9.23	6.40	5.28	4.66	4.26	3.98	3.77	3.61	3.48
30	.100	2.88	2.49	2.28	2.14	2.05	1.98	1.93	1.88	1.85
	.050	4.17	3.32	2.92	2.69	2.53	2.42	2.33	2.27	2.21
	.025	5.57	4.18	3.59	3.25	3.03	2.87	2.75	2.65	2.57
	.010	7.56	5.39	4.51	4.02	3.70	3.47	3.30	3.17	3.07
	.005	9.18	6.35	5.24	4.62	4.23	3.95	3.74	3.58	3.45
40	.100	2.84	2.44	2.23	2.09	2.00	1.93	1.87	1.83	1.79
	.050	4.08	3.23	2.84	2.61	2.45	2.34	2.25	2.18	2.12
	.025	5.42	4.05	3.46	3.13	2.90	2.74	2.62	2.53	2.45
	.010	7.31	5.18	4.31	3.83	3.51	3.29	3.12	2.99	2.89
	.005	8.83	6.07	4.98	4.37	3.99	3.71	3.51	3.35	3.22
60	.100	2.79	2.39	2.18	2.04	1.95	1.87	1.82	1.77	1.74
	.050	4.00	3.15	2.76	2.53	2.37	2.25	2.17	2.10	2.04
	.025	5.29	3.93	3.34	3.01	2.79	2.63	2.51	2.41	2.33
	.010	7.08	4.98	4.13	3.65	3.34	3.12	2.95	2.82	2.72
	.005	8.49	5.79	4.73	4.14	3.76	3.49	3.29	3.13	3.01
120	.100	2.75	2.35	2.13	1.99	1.90	1.82	1.77	1.72	1.68
	.050	3.92	3.07	2.68	2.45	2.29	2.17	2.09	2.02	1.96
	.025	5.15	3.80	3.23	2.89	2.67	2.52	2.39	2.30	2.22
	.010	6.85	4.79	3.95	3.48	3.17	2.96	2.79	2.66	2.56
	.005	8.18	5.54	4.50	3.92	3.55	3.28	3.09	2.93	2.81
∞	.100	2.71	2.30	2.08	1.94	1.85	1.77	1.72	1.67	1.63
	.050	3.84	3.00	2.60	2.37	2.21	2.10	2.01	1.94	1.88
	.025	5.02	3.69	3.12	2.79	2.57	2.41	2.29	2.19	2.11
	.010	6.63	4.61	3.78	3.32	3.02	2.80	2.64	2.51	2.41
	.005	7.88	5.30	4.28	3.72	3.35	3.09	2.90	2.74	2.62

10	12	15	20	24	30	40	60	120	∞	p	f₂
				Numerator df (f_1)							
1.83	1.78	1.73	1.68	1.65	1.62	1.58	1.55	1.51	1.47	.100	29
2.18	2.10	2.03	1.94	1.90	1.85	1.81	1.75	1.70	1.64	.050	
2,53	2,43	2.32	2.21	2.15	2.09	2.03	1.96	1.89	1.81	.025	
3.00	2.87	2.73	2.57	2.49	2.41	2.33	2.23	2.14	2.03	.010	
3.38	3.21	3.04	2.86	2.76	2.66	2.56	2.45	2.33	2.21	.005	
1.82	1.77	1.72	1.67	1.64	1.61	1.57	1.54	1.50	1.46	.100	30
2.16	2.09	2.01	1.93	1.89	1.84	1.79	1.74	1.68	1.62	.050	
2.51	2.41	2.31	2.20	2.14	2.07	2.01	1.94	1.87	1.79	.025	
2.98	2.84	2.70	2.55	2.47	2.39	2.30	2.21	2.11	2.01	.010	
3.34	3.18	3.01	2.82	2.73	2.63	2.52	2.42	2.30	2.18	.005	
1.76	1.71	1.66	1.61	1.57	1.54	1.51	1.47	1.42	1.38	.100	40
2.08	2.00	1.92	1.84	1.79	1.74	1.69	1.64	1.58	1.51	.050	
2.39	2.29	2.18	2.07	2.01	1.94	1.88	1.80	1.72	1.64	.025	
2.80	2.66	2.52	2.37	2.29	2.20	2.11	2.02	1.92	1.80	.010	
3.12	2.95	2.78	2.60	2.50	2.40	2.30	2.18	2.06	1.93	.005	
1.71	1.66	1.60	1.54	1.51	1.48	1.44	1.40	1.35	1.29	.100	60
1.99	1.92	1.84	1.75	1.70	1.65	1.59	1.53	1.47	1.39	.050	
2.27	2.17	2.06	1.94	1.88	1.82	1.74	1.67	1.58	1.48	.025	
2.63	2.50	2.35	2.20	2.12	2.03	1.94	1.84	1.73	1.60	.010	
2.90	2.74	2.57	2.39	2.29	2.19	2.08	1.96	1.83	1.69	.005	
1.65	1.60	1.55	1.48	1.45	1.41	1.37	1.32	1.26	1.19	.100	120
1.91	1.83	1.75	1.66	1.61	1.55	1.50	1.43	1.35	1.25	.050	
2.16	2.05	1.94	1.82	1.76	1.69	1.61	1.53	1.43	1.31	.025	
2.47	2.34	2.19	2.03	1.95	1.86	1.76	1.66	1.53	1.38	.010	
2.71	2.54	2.37	2.19	2.09	1.98	1.87	1.75	1.61	1.43	.005	
1.60	1.55	1.49	1.42	1.38	1.34	1.30	1.24	1.17	1.00	.100	∞
1.83	1.75	1.67	1.57	1.52	1.46	1.39	1.32	1.22	1.00	.050	
2.05	1.94	1.83	1.71	1.64	1.57	1.48	1.39	1.27	1.00	.025	
2.32	2.18	2.04	1.88	1.79	1.70	1.59	1.47	1.32	1.00	.010	
2.52	2.36	2.19	2.00	1.90	1.79	1.67	1.53	1.36	1.00	.005	

SOURCE: A portion of "Tables of percentage points of the inverted beta (*F*) distribution," *Biometrika*, **33** (1943) by M. Merrington and C. M. Thompson and from Table 18 of *Biometrika Tables for Statisticians*, Vol. 1, Cambridge University Press, 1954, edited by E. S. Pearson and H. O. Hartley. Reproduced with permission of the authors, editors, and *Biometrika* trustees.

Table A6. Studentized Range, q-Variate Probabilities
(α in column 2, q_α in body)

f = df for s^2	α	\multicolumn{10}{c}{a = Number of Group Means}									
		2	3	4	5	6	7	8	9	10	11
5	.05	3.64	4.60	5.22	5.67	6.03	6.33	6.58	6.80	6.99	7.17
	.01	5.70	6.97	7.80	8.42	8.91	9.32	9.67	9.97	10.24	10.48
6	.05	3.46	4.34	4.90	5.31	5.63	5.89	6.12	6.32	6.49	6.65
	.01	5.24	6.33	7.03	7.56	7.97	8.32	8.61	8.87	9.10	9.30
7	.05	3.34	4.16	4.68	5.06	5.36	5.61	5.82	6.00	6.16	6.30
	.01	4.95	5.92	6.54	7.01	7.37	7.68	7.94	8.17	8.37	8.55
8	.05	3.26	4.04	4.53	4.89	5.17	5.40	5.60	5.77	5.92	6.05
	.01	4.74	5.63	6.20	6.63	6.96	7.24	7.47	7.68	7.87	8.03
9	.05	3.20	3.95	4.42	4.76	5.02	5.24	5.43	5.60	5.74	5.87
	.01	4.60	5.43	5.96	6.35	6.66	6.91	7.13	7.32	7.49	7.65
10	.05	3.15	3.88	4.33	4.65	4.91	5.12	5.30	5.46	5.60	5.72
	.01	4.48	5.27	5.77	6.14	6.43	6.67	6.87	7.05	7.21	7.36
11	.05	3.11	3.82	4.26	4.57	4.82	5.03	5.20	5.35	5.49	5.61
	.01	4.39	5.14	5.62	5.97	6.25	6.48	6.67	6.84	6.99	7.13
12	.05	3.08	3.77	4.20	4.51	4.75	4.95	5.12	5.27	5.40	5.51
	.01	4.32	5.04	5.50	5.84	6.10	6.32	6.51	6.67	6.81	6.94
13	.05	3.06	3.73	4.15	4.45	4.69	4.88	5.05	5.19	5.32	5.43
	.01	4.26	4.96	5.40	5.73	5.98	6.19	6.37	6.53	6.67	6.79
14	.05	3.03	3.70	4.11	4.41	4.64	4.83	4.99	5.13	5.25	5.36
	.01	4.21	4.89	5.32	5.63	5.88	6.08	6.26	6.41	6.54	6.66
15	.05	3.01	3.67	4.08	4.37	4.60	4.78	4.94	5.08	5.20	5.31
	.01	4.17	4.83	5.25	5.56	5.80	5.99	6.16	6.31	6.44	6.55
16	.05	3.00	3.65	4.05	4.33	4.56	4.74	4.90	5.03	5.15	5.26
	.01	4.13	4.78	5.19	5.49	5.72	5.92	6.08	6.22	6.35	6.46
17	.05	2.98	3.63	4.02	4.30	4.52	4.71	4.86	4.99	5.11	5.21
	.01	4.10	4.74	5.14	5.43	5.66	5.85	6.01	6.15	6.27	6.38
18	.05	2.97	3.61	4.00	4.28	4.49	4.67	4.82	4.96	5.07	5.17
	.01	4.07	4.70	5.09	5.38	5.60	5.79	5.94	6.08	6.20	6.31
19	.05	2.96	3.59	3.98	4.25	4.47	4.65	4.79	4.92	5.04	5.14
	.01	4.05	4.67	5.05	5.33	5.55	5.73	5.89	6.02	6.14	6.25
20	.05	2.95	3.58	3.96	4.23	4.45	4.62	4.77	4.90	5.01	5.11
	.01	4.02	4.64	5.02	5.29	5.51	5.69	5.84	5.97	6.09	6.19
24	.05	2.92	3.53	3.90	4.17	4.37	4.54	4.68	4.81	4.92	5.01
	.01	3.96	4.54	4.91	5.17	5.37	5.54	5.69	5.81	5.92	6.02
30	.05	2.89	3.49	3.84	4.10	4.30	4.46	4.60	4.72	4.83	4.92
	.01	3.89	4.45	4.80	5.05	5.24	5.40	5.54	5.65	5.76	5.85
40	.05	2.86	3.44	3.79	4.04	4.23	4.39	4.52	4.63	4.74	4.82
	.01	3.82	4.37	4.70	4.93	5.11	5.27	5.39	5.50	5.60	5.69
60	.05	2.83	3.40	3.74	3.98	4.16	4.31	4.44	4.55	4.65	4.73
	.01	3.76	4.28	4.60	4.82	4.99	5.13	5.25	5.36	5.45	5.53
120	.05	2.80	3.36	3.69	3.92	4.10	4.24	4.36	4.48	4.56	4.64
	.01	3.70	4.20	4.50	4.71	4.87	5.01	5.12	5.21	5.30	5.38
∞	.05	2.77	3.31	3.63	3.86	4.03	4.17	4.29	4.39	4.47	4.55
	.01	3.64	4.12	4.40	4.60	4.76	4.88	4.99	5.08	5.16	5.23

SOURCE: This table is abridged from Table 29, *Biometrika Tables for Statisticians*, vol. 1, Cambridge University Press, 1954. It is reproduced with permission of the *Biometrika* trustees and the editors, E. S. Pearson and H. O. Hartley. The original work appeared in a paper by J. M. May, Extended and corrected tables of the upper percentage points of the "Studentized" range, *Biometrika*, **39**, 192–193 (1952).

Table A6. (*Continued*)

12	13	14	15	16	17	18	19	20	α	$f = $ df for s^2
7.32	7.47	7.60	7.72	7.83	7.93	8.03	8.12	8.21	.05	5
10.70	10.89	11.08	11.24	11.40	11.55	11.68	11.81	11.93	.01	
6.79	6.92	7.03	7.14	7.24	7.34	7.43	7.51	7.59	.05	6
9.49	9.65	9.81	9.95	10.08	10.21	10.32	10.43	10.54	.01	
6.43	6.55	6.66	6.76	6.85	6.94	7.02	7.09	7.17	.05	7
8.71	8.86	9.00	9.12	9.24	9.35	9.46	9.55	9.65	.01	
6.18	6.29	6.39	6.48	6.57	6.65	6.73	6.80	6.87	.05	8
8.18	8.31	8.44	8.55	8.66	8.76	8.85	8.94	9.03	.01	
5.98	6.09	6.19	6.28	6.36	6.44	6.51	6.58	6.64	.05	9
7.78	7.91	8.03	8.13	8.23	8.32	8.41	8.49	8.57	.01	
5.83	5.93	6.03	6.11	6.20	6.27	6.34	6.40	6.47	.05	10
7.48	7.60	7.71	7.81	7.91	7.99	8.07	8.15	8.22	.01	
5.71	5.81	5.90	5.99	6.06	6.14	6.20	6.26	6.33	.05	11
7.25	7.36	7.46	7.56	7.65	7.73	7.81	7.88	7.95	.01	
5.62	5.71	5.80	5.88	5.95	6.03	6.09	6.15	6.21	.05	12
7.06	7.17	7.26	7.36	7.44	7.52	7.59	7.66	7.73	.01	
5.53	5.63	5.71	5.79	5.86	5.93	6.00	6.05	6.11	.05	13
6.90	7.01	7.10	7.19	7.27	7.34	7.42	7.48	7.55	.01	
5.46	5.55	5.64	5.72	5.79	5.85	5.92	5.97	6.03	.05	14
6.77	6.87	6.96	7.05	7.12	7.20	7.27	7.33	7.39	.01	
5.40	5.49	5.58	5.65	5.72	5.79	5.85	5.90	5.96	.05	15
6.66	6.76	6.84	6.93	7.00	7.07	7.14	7.20	7.26	.01	
5.35	5.44	5.52	5.59	5.66	5.72	5.79	5.84	5.90	.05	16
6.56	6.66	6.74	6.82	6.90	6.97	7.03	7.09	7.15	.01	
5.31	5.39	5.47	5.55	5.61	5.68	5.74	5.79	5.84	.05	17
6.48	6.57	6.66	6.73	6.80	6.87	6.94	7.00	7.05	.01	
5.27	5.35	5.43	5.50	5.57	5.63	5.69	5.74	5.79	.05	18
6.41	6.50	6.58	6.65	6.72	6.79	6.85	6.91	6.96	.01	
5.23	5.32	5.39	5.46	5.53	5.59	5.65	5.70	5.75	.05	19
6.34	6.43	6.51	6.58	6.65	6.72	6.78	6.84	6.89	.01	
5.20	5.28	5.36	5.43	5.49	5.55	5.61	5.66	5.71	.05	20
6.29	6.37	6.45	6.52	6.59	6.65	6.71	6.76	6.82	.01	
5.10	5.18	5.25	5.32	5.38	5.44	5.50	5.54	5.59	.05	24
6.11	6.19	6.26	6.33	6.39	6.45	6.51	6.56	6.61	.01	
5.00	5.08	5.15	5.21	5.27	5.33	5.38	5.43	5.48	.05	30
5.93	6.01	6.08	6.14	6.20	6.26	6.31	6.36	6.41	.01	
4.91	4.98	5.05	5.11	5.16	5.22	5.27	5.31	5.36	.05	40
5.77	5.84	5.90	5.96	6.02	6.07	6.12	6.17	6.21	.01	
4.81	4.88	4.94	5.00	5.06	5.11	5.16	5.20	5.24	.05	60
5.60	5.67	5.73	5.79	5.84	5.89	5.93	5.98	6.02	.01	
4.72	4.78	4.84	4.90	4.95	5.00	5.05	5.09	5.13	.05	120
5.44	5.51	5.56	5.61	5.66	5.71	5.75	5.79	5.83	.01	
4.62	4.68	4.74	4.80	4.85	4.89	4.93	4.97	5.01	.05	∞
5.29	5.35	5.40	5.45	5.49	5.54	5.57	5.61	5.65	.01	

$a = $ Number of Group Means

Index

Applied Probability and Statistics (Continued)

JUDGE, HILL, GRIFFITHS, LÜTKEPOHL and LEE • Introduction to the Theory and Practice of Econometrics

JUDGE, GRIFFITHS, HILL, LÜTKEPOHL and LEE • The Theory and Practice of Econometrics, *Second Edition*

KALBFLEISCH and PRENTICE • The Statistical Analysis of Failure Time Data

KISH • Survey Sampling

KUH, NEESE, and HOLLINGER • Structural Sensitivity in Econometric Models

KEENEY and RAIFFA • Decisions with Multiple Objectives

LAWLESS • Statistical Models and Methods for Lifetime Data

LEAMER • Specification Searches: Ad Hoc Inference with Nonexperimental Data

LEBART, MORINEAU, and WARWICK • Multivariate Descriptive Statistical Analysis: Correspondence Analysis and Related Techniques for Large Matrices

LINHART and ZUCCHINI • Model Selection

LITTLE and RUBIN • Statistical Analysis with Missing Data

McNEIL • Interactive Data Analysis

MAINDONALD • Statistical Computation

MANN, SCHAFER and SINGPURWALLA • Methods for Statistical Analysis of Reliability and Life Data

MARTZ and WALLER • Bayesian Reliability Analysis

MIKÉ and STANLEY • Statistics in Medical Research: Methods and Issues with Applications in Cancer Research

MILLER • Beyond ANOVA, Basics of Applied Statistics

MILLER • Survival Analysis

MILLER, EFRON, BROWN, and MOSES • Biostatistics Casebook

MONTGOMERY and PECK • Introduction to Linear Regression Analysis

NELSON • Applied Life Data Analysis

OSBORNE • Finite Algorithms in Optimization and Data Analysis

OTNES and ENOCHSON • Applied Time Series Analysis: Volume I, Basic Techniques

OTNES and ENOCHSON • Digital Time Series Analysis

PANKRATZ • Forecasting with Univariate Box-Jenkins Models: Concepts and Cases

PIELOU • Interpretation of Ecological Data: A Primer on Classification and Ordination

PLATEK, RAO, SARNDAL and SINGH • Small Area Statistics: An International Symposium

POLLOCK • The Algebra of Econometrics

PRENTER • Splines and Variational Methods

RAO and MITRA • Generalized Inverse of Matrices and Its Applications

RÉNYI • A Diary on Information Theory

RIPLEY • Spatial Statistics

RIPLEY • Stochastic Simulation

RUBIN • Multiple Imputation for Nonresponse in Surveys

RUBINSTEIN • Monte Carlo Optimization, Simulation, and Sensitivity of Queueing Networks

SCHUSS • Theory and Applications of Stochastic Differential Equations

SEAL • Survival Probabilities: The Goal of Risk Theory

SEARLE • Linear Models

SEARLE • Matrix Algebra Useful for Statistics

SPRINGER • The Algebra of Random Variables

STEUER • Multiple Criteria Optimization

STOYAN • Comparison Methods for Queues and Other Stochastic Models

TIJMS • Stochastic Modeling and Analysis: A Computational Approach

TITTERINGTON, SMITH, and MAKOV • Statistical Analysis of Finite Mixture Distributions

UPTON • The Analysis of Cross-Tabulated Data

UPTON and FINGLETON • Spatial Data Analysis by Example, Volume I: Point Pattern and Quantitative Data